Food Processing and Preservation

Edited by Roua Lajnaf

Published in London, United Kingdom

Food Processing and Preservation
http://dx.doi.org/10.5772/intechopen.106118
Edited by Roua Lajnaf

Contributors
Sadoth Sandoval Torres, Larissa Giovana Reyes López, Lilia Leticia Méndez Lagunas, Luis Gerardo Barriada Bernal, Juan Rodríguez Ramírez, Samuel Tunde Olorunsogo, Bolanle Adenike Adejumo, M A Firdaus, Sunita Mishra, Olugbenga Ben Ogunmoyela, Musa Omotayo Jimoh, Roua Lajnaf, Hamadi Attia, Mohamed Ali Ayadi

First published in London, United Kingdom, 2023 by IntechOpen
IntechOpen is the global imprint of INTECHOPEN LIMITED, registered in England and Wales, registration number: 11086078, 5 Princes Gate Court, London, SW7 2QJ, United Kingdom

British Library Cataloguing-in-Publication Data
A catalogue record for this book is available from the British Library

Additional hard and PDF copies can be obtained from orders@intechopen.com

Food Processing and Preservation
Edited by Roua Lajnaf
p. cm.
Print ISBN 978-1-83768-688-9
Online ISBN 978-1-83768-689-6
eBook (PDF) ISBN 978-1-83768-690-2

Meet the editor

Dr. Roua Lajnaf, a scientist from the National Engineering School of Sfax, Tunisia, and the University of Montpellier, France, has more than 11 years of experience in various fields of dairy science. Her areas of expertise include the purification of milk proteins and the study of the effect of different processes (heat treatment, high pressure, acidification, fermentation) on their properties, including techno-functional properties and allergenicity. She has worked in academics as well as in food bioprocessing. Dr. Lajnaf has published numerous research papers, review articles, book chapters, and national patents on different subjects in food science, especially dairy biochemistry and processing. She is also a reviewer for various journals in food biochemistry and physical chemistry. Dr. Lajnaf is currently an assistant professor at the University of Monastir, Tunisia. She is the recipient of the 2021 L'Oréal-UNESCO For Women in Science Award.

Contents

Preface

Food Processing and Preservation describes the latest research and knowledge about food processing and preservation technologies. Food processing and preservation is a major concern for consumers and thus have attracted much attention. They can involve one or a combination of various processes during the manufacturing of foods, including washing, chopping, pasteurizing, roasting, freezing, fermenting, packaging, cooking, and many more.

Food processing and preservation can have many beneficial effects to obtain various food products and to stabilize foods against alteration. In recent years, consumer preferences have led to increased demand for high-quality foods in terms of nutritional physiology, leading to increased interest in fresh food products. As such, researchers and scientists have been working to improve the sustainability and quality of food products using new technologies and processes.

This book is an authoritative source of information about food processing and preservation technologies. Chapters address such topics as different preservation processes, the effect of these processes on the quality of food products, new perspectives on established processes, and innovative and emerging technologies in food processing.

This book is the result of the combined efforts of experts in food processing from around the world who are affiliated with industry, research institutions, and academia. It is unique both in depth and breadth and is a useful reference for professionals in industry and academia as well as students studying food science and food processing. We wish to thank all the authors for their excellent contributions and patience throughout the publication process.

Dr. Roua Lajnaf
Alimentary Analysis Unit,
National Engineering School of Sfax,
Sfax, Tunisia

Montpellier University,
UMR IATE,
Montpellier, France

Section 1

Introduction

Chapter 1

Introductory Chapter: Novel Thermal and Non-Thermal Technologies for Food Processing

Roua Lajnaf

1. Introduction

Food Processing is the set of techniques and methods used to transform raw materials into final products. Quality of foods is a great concern when processing methods are used for food preservation. Thus, food preservation whilst ensuring its quality and safety has been a prime goal of food processors which is mainly attributed to the preferences of the consumer which increased the demand for high quality foods in terms of nutritional physiology and technological quality. Since prehistoric times, processing of food materials including sun drying, salting, fermenting and smoking have been used to preserve foodstuffs in order to make them edible [1]. Indeed, food processing methods have been developed to inactivate pathogenic bacteria, toxins as well as detrimental constituents so that the obtained processed foods can meet safety and shelf-stability requirements to the consumer. Nowadays, more emphasis was attributed on informatively labeled, high-quality and value-added foods which are convenient to use beyond the traditional requirements [2]. In the recent years, the awareness of foods that are beneficial to consumer health has increased, especially about the effects of processing on the functional components of the various food products. Consequently, a number of health-conscious consumers try to more eat raw foods. However, many foods are available only when they are processed, furthermore, they need to be processed to make them safe for consumption [1]. The purpose of this introductive chapter is to provide a general perspective of the novel thermal and non-thermal processing technologies currently available in connection with their efficiency and their impact once implemented by the food industry. Food processing are commonly divided into two main types: thermal and non-thermal processing as shown in **Figure 1** [1, 3, 4]. Thermal processing is the most common and traditional technique because of its ability and efficiency to inactivate microorganisms as well as spoilage enzymes [5]. However, severe heat treatments may induce chemical and physical changes to food products. These techniques can even induce the formation of toxic compounds and can reduce the bioavailability of some importants nutrients. Furthermore, an adverse effect on the sensory properties of foods caused by thermal processing has been reported [1]. As a consequence, softer processing techniques including novel thermal and non-thermal processing has become the new trend. Indeed, these processing techniques are claimed to combine the high quality of industrial foods and the improved functionality.

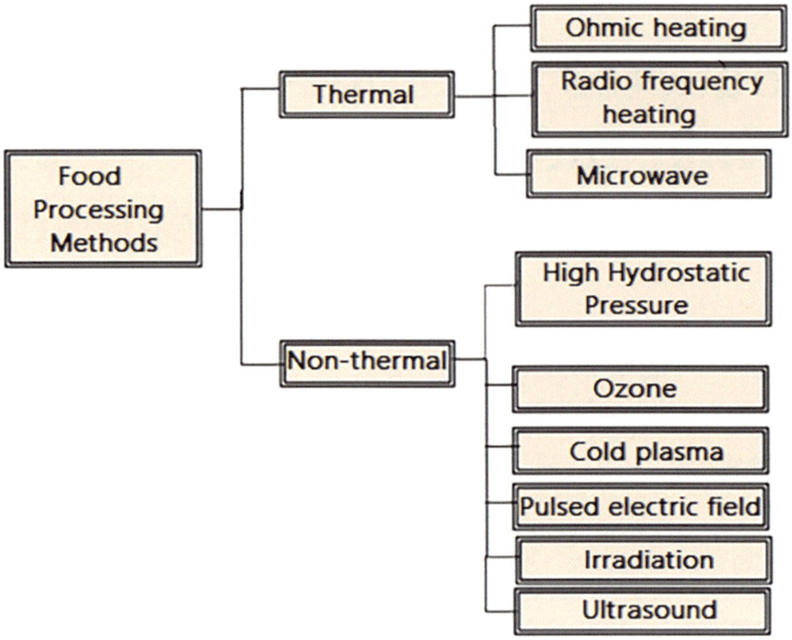

Figure 1.
Thermal and non-thermal food processing methods.

They were also reported to be more cost efficient and environmentally friendly compared to classic thermal processing techniques. Therefore, the aim of this introductory chapter is to provide a critical review of novel food preservation processes including both thermal and non-thermal technologies.

2. Thermal food processing methods

Overall, conventional food preservation processes consist of exposing food to a very high temperature, leading to the reduction of the contamination or microbial load from food.

Thermal processing techniques persists as the most widely used techniques for prolonging the shelf life of foods and ensuring their microbiological safety [1]. However, these processes result some undesirable changes in food including the loss of nutritional components which are temperature-sensitive, the change in the texture of food and in the organoleptic characteristics of treated foods [6]. The thermal treatment used for preservation result also in the formation of chemical toxicants in food which are carcinogenic and harmful to the human body. However, the amount and the type these toxicants formed depend on the treated food and the type of thermal method used for cooking food [7, 8].

Furthermore, the functional benefits of thermally processed foods are still doubtful as thermal treatment causes considerable changes in the nutritional attributes and the quality of foods [9]. For instance, vitamins are among the most sensitive components in foods to be affected by thermal treatments. Heat-sensitive vitamins include both of fat-soluble vitamins such as vitamins A (in the presence of oxygen), D, E and β-carotene, and water-soluble vitamins such as vitamin C (ascorbic acid),

B1 (thiamine), B2 (riboflavin) in an acid environment, nicotinic acid, biotin C and pantothenic acid [10]. Novel thermal processing techniques are characterized by novel heating alternatives that can offer quicker heating rates leading to minimization of nutrient degradation and adverse reactions. These techniques include ohmic, radio frequency and microwave heating.

2.1 Ohmic heating

Ohmic heating is generally included to the group of novel processing technologies and especially to the novel alternatives to conventional thermal processing. This technology has been known since the nineteenth century as it was applied to pasteurize milk [11, 12]. Ohmic heating is direct electro-heating where electrical current is applied directly to the food while microwave processing and radio frequency heating are indirect electro-heating where the electrical energy is firstly converted to electromagnetic radiation that subsequently generates heat within a product [13]. Various ohmic heating processes that use electrical current for heating food have been used and developed. This technology has been also widely studied and reviewed in many scientific publications [11, 14–16].

Generally, most of scientific publications that studied the effect of ohmic heating processes on foods were carried out at either 20 kHz or 50 kHz [17–19]. For instance, ohmic heating frequency at 10 or 60 kHz on the inactivation kinetics of *Geobacillus stearothermophilus* spores showed an improved antimicrobial efficiency at higher temperatures as reported by Somavat et al. [20, 21]. Most commonly studied food matrices were meat, processed meat products and liquid foods including fruit, vegetable juices, milk and milk-analogues. Indeed, the main drawback for this technology is that ohmic treatments can alter the textural properties of foods despite a correct optimization of the ohmic process conditions [12, 16].

2.2 Radio frequency heating

Radio frequency heating process involves the direct transfer of electromagnetic energy into food product, thus, it induced volumetric heating due to frictional interaction between different molecules [22]. The allowed frequencies for the applications of this technique are 13.56, 27.12, and 40.68 MHz [22]. The greater wavelength at radio frequencies compared to those of microwave heating justifies the significant advantages of radio frequencies over microwaves, especially in the case of food processing applications [23]. Radio frequencies heating presents similar advantages to ohmic and microwave processes when compared with conventional heat-processing technologies. Some specific advantages of radio frequencies heating over those alternative volumetric technologies are noted. However, radio frequencies processing presents some disadvantages which are: the higher equipment and operating costs when compared to conventional heating systems, the reduced power density when compared to microwave heating, and the limited research efforts regarding the determination of food treated by radio frequencies dielectric properties [23, 24].

2.3 Microwave processing

Microwave technique is considered as a novel thermal treatment whose used have increased over the last years either in food industry or for domestic use. Domestic microwaves generally operate at a frequency of 2.45GHz while industrial microwave systems operate at frequencies that range between 915 MHz and 2.45GHz.The microwave are distinguished by generating heat instantly which significantly reduces the processing time and operational cost when compared with the conventional dry-heating methods [3, 25].

Overall, microwave heating is used in both of pasteurization and sterilization. Indeed, pasteurization is a process in which only pathogenic microorganisms in the vegetative form are destroyed by thermal treatment in order to enhance food safety and shelf life. For microwave processing, the destruction of microbes at sub-lethal temperatures was attributed to the selective heating, cell membrane rupture, electroporation, and magnetic field coupling [26].

The main advantages of microwaves processing are the less time of the process and the fast and efficient heating. Indeed, product quality and food nutritional and sensory qualities are improved with reduced drying time [3]. However, microwaves also show disadvantages, such as degradation of treated food products by dry heating and food dehydration. However, one of the main disadvantages recognized in microwave heating is the non-uniform temperature distribution resulting in hot and cold spots in microwave-heated products [12].

3. Non-thermal food processing methods

Recent developments in food preservation processes operations involve novel technologies that minimize the deleterious effects of heat on the nutritional and sensory properties of foods. Promising methods include non-thermal processing which can be conducted at ambient or slightly above-ambient temperature and causes no change in the nutirional composition of food and the texture which remaines intact [1, 4].

Since the last few decades, various non-thermal technology for food processing treatments came into light including pulsed electric field, ultrasonication, cold plasma etc. These non-thermal treatments result in the reduction of the microbial load in the treated food with an increase in shelf-life and with good sensory and textural characteristics as they unmask food to treatment conditions for a fraction of seconds [27]. Furthermore, the preservation effects of non-thermal technologies are more than those of thermal technologies because there is no risk for the formation of any toxic or undesirable products in foods since it is not exposed to high temperatures [4].

3.1 Ultrasonication processing

Ultrasonication is an emerging non-thermal technology in food industry, whereas it is used in other processing sectors [28]. It refers to sound waves at a higher frequency than that of human hearing (between 20 kHz and 10 MHz). The major effects of ultrasound on liquid systems is contributed by cavitation phenomena, which is the physical processes that create and implode micro-bubbles of gases dissolved in a liquid by the compression and decompression of the different molecules that constitute the liquid medium [1]. Ultrasonication is generally used with different frequencies including low-frequency, medium-frequency, and highfrequency ultrasonication, with frequencies that range of 20 kHz–100 kHz, 100 kHz – 1 MHz, and 1 MHz–100 MHz, respectively [29]. Ultrasonication is useful for the process of degassing in carbonated beverages. It is also a good replacement for the processes of preservation including pasteurization and sterilization in order to reduce the microbial load in both of foods and food products [30]. Ultrasound has successfully proven its usefulness in the food sector in various areas such as food preservation, extraction, intensified synthesis, and improvement of the physical and chemical properties of food [4]. However, this treatment must be studied on bulk food in order to understand its effect so that it can be implemented at industrial scale.

3.2 High hydrostatic pressure processing

Application of High hydrostatic pressure mainly deals with pressure as a preservation method with great potential to produce microbiologically safer food products. During this process, the pressure ranging between 100 and 600 MPa is transmitted uniformly and instantly, with a little variation in temperature upon increasing pressure regardless of the size of the food (the rate of temperature increase is about 3°C/100 MPa) [31].

This method causes microbial cell injury and does not alter low-energy covalent bonds. As the covalent bonds have low compressibility and would not break within the ranges of high pressure used in food processing, the primary structure of molecules in food such as proteins or fats remains intact [32]. High hydrostatic pressure can bring about a significant decimal decrease in the population of pathogenic Gram-positive bacteria and Gram-negative bacteria, furthermore, it helps in food preservation for a longer duration. However, the reduction in microbial load depends greatly on the pressure, temperature during treatment and type of food processed [4]. The quality of High hydrostatic pressure-processed food in terms of nutritional components, sensory, and texture is excellent due to the short period of food exposure to treatment conditions [33].

3.3 Radiation processing

Radiation processing of food is classified as a physical and non-thermal mode of food preservation, hence, it is called cold pasteurization [1]. Food produces can be exposed to either ionizing radiation or non-ionizing radiation in order to destroy pathigenic microorganisms or insects in the food. Ionizing radiation is generated by either electron beams, X-rays or gamma rays leading to the inactivation of microorganisms by damaging their DNA, while non-ionizing radiation is generated from ultraviolet rays, visible light, microwaves or infrared. Applications radiation processing are mostly employed in the food processing sector especially for the preservation of food products. This technique is effective against pathogenic microbes including *E. coli*, *Salmonella* and *Staphylococcus* [34].

However, the use of radiation processing result in some undesirable changes in food if treated at high irradiation doses. For instance, the color of meat as well its lipids have been slightly changed a which may lead to rejection by consumers. Therefore, irradiation is usually done with a low dose with the combination of the use of antimicrobial agents in order to achieve the desired inactivation in food with no change in the food composition and processed food products [35].

3.4 Pulsed electric field processing

Pulsed electric field technology is classified as a non-thermal technology for food processing which is capable to inactivate microorganisms and enzymes and to retain health-related compounds concurrently [1]. The Pulsed electric field process is commonly applied to liquid foods by the application of a series of short and high-voltage pulses (25–80 kV/cm) to a liquid food [1]. Since food is exposed to pulsed electric field for a very short duration of time ranging between few milliseconds to nanosecond, there is no chance of heating and hence, undesirable changes in food due to high temperature are eliminated [36].

The mechanism of pulsed electric field causing microorganism inactivation is known as electroporation of cells. Indeed, pulsed electric field causes damage to the cell membrane of microbes through tension in the cell membrane attributable to electromechanical compression which facilitates the formation of pores in the membrane [37].

The efficiency of this process in reducing microbial load depends on the intensity of field applied, temperature, the total exposure time, and energy [4]. Previous studies have shown that pulsed electric field process was effective against *E. coli* in flowable food like pineapple and orange juice and coconut water with an intensity of 5.6 W/cm^2 [38]. Similar results reported that pulsed electric field intensity are also effective for microbial inactivation in fruit juices [39]. Apart from microbial inactivation, pulsed electric field process is also effective in the deactivation of food spoilage enzymes [4]. For instance, the most common discussed enzymes are polyphenoloxidase and peroxidase which catalyze oxidation of phenolic compounds leading to enzymatic browning of vegetables [1].

3.5 Ozone

Ozone is generally employed as an effective antibacterial agent against many bacteria in food in gas form or in ozonated water as it can be mixed with water to form [4]. There are many mechanisms by which ozone causes microbial cell death. For instance, ozone causes damages of the microbial cell membranes as it alters the permeability of cells. Furthermore, ozone is known to damage the structure of proteins that leads to the malfunctioning of microbial enzymes, which results in microbial cell death [40]. Ozone showed its effectiveness of ozone against Listeria monocytogenes present in meat by a treatment of 280 mg O_3/m^3 for 5 h with pulse of ozone passed after 10 min for 30 min duration. Furthermore, ozone was efficient for the inactivation of *Salmonella* and spoilage microorganisms [4, 41, 42]. Ozone treatment is also effective in the inactivation of toxins present in food [40]. However, this reactive molecule reacts with many components in food which could induce undesirable changes. It also induces oxidation in food lipids. Further studies are needed in order to reduce these undesirable changes in food and to improve its acceptability.

3.6 Cold plasma technology

In the food industry, cold plasma can be used for the reduction of the microbial load in food products or on the surface of food which enhances the physical and chemical properties of different food constituents including lipids and proteins. Thus, it is used for the sterilization of food processing equipment, treatment of food packaging material, inactivation of food spoilage enzymes, and treatment of wastewater [43]. There are no risk of thermal damage to heat-sensitive food material as the used temperature is ambient. Cold nitrogen plasma shows significant inhibitory action on *Salmonella enterica* after a treatment of 600 W for 2 min [44]. Furthermore, a reduction of 97.9% and 99.3% in the growth of fungal species such as *Aspergillus parasiticus* and *Aspergillus flavus*, respectively, after a treatment at 60 W plasma power on the ground nut surface [45].

Author details

Roua Lajnaf[1,2]

1 Alimentary Analysis Unit, National Engineering School of Sfax, Sfax, Tunisia

2 Montpellier University, UMR IATE, Montpellier, France

*Address all correspondence to: roua_lajnaf@yahoo.fr

References

[1] Jongyingcharoen JS, Ahmad I. Chapter 11: Thermal and Non-Thermal Processing of Functional Foods, Functional foods and dietary supplements: Processing effects and health benefits. Wiley; 2014. pp. 295-324

[2] Awuah GB, Ramaswamy HS, Economides A. Thermal processing and quality: Principles and overview. Chemical Engineering and Processing Process Intensification. 2007;**46**:584-602

[3] Vapor A, Mendonça A, Tomaz CT. Processes for reducing egg allergenicity: Advances and different approaches. Food Chemistry. 2022;**367**:130568

[4] Jadhav HB, Annapure US, Deshmukh RR. Non-thermal technologies for food processing. Frontiers in Nutrition. 2021;**8**:657090

[5] Rawson A, Patras A, Tiwari BK, Noci F, Koutchma T, Brunton N. Effect of thermal and non thermal processing technologies on the bioactive content of exotic fruits and their products: Review of recent advances. Food Research International. 2011;**44**:1875-1887

[6] Hernández-Hernández HM, Moreno-Vilet L, Villanueva-Rodríguez SJ. Current status of emerging food processing technologies in Latin America: Novel non-thermal processing. Innovative Food Science and Emerging Technologies. 2019;**58**:102233

[7] Oz E. Effects of smoke flavoring using different wood chips and barbecuing on the formation of polycyclic aromatic hydrocarbons and heterocyclic aromatic amines in salmon fillets. PLoS One. 2020;**15**:0227508

[8] Oz F, Kızıl M, Çelık T. Effects of different cooking methods on the formation of heterocyclic aromatic amines in goose meat. Journal of Food Processing & Preservation. 2016;**40**:1047-1053

[9] Tewari G, Juneja V. Advances in Thermal and Non-thermal Food Preservation. Ames, IA: Blackwell Publishing; 2008. pp. 1-281

[10] Ryley J, Kajda P. Vitamins in thermal processing. Food Chemistry. 1994;**49**:119-129

[11] Cappato LP, Ferreira MVS, Guimaraes JT, Portela JB, Costa ALR, Freitas MQ, et al. Ohmic heating in dairy processing: Relevant aspects for safety and quality. Trends in Food Science and Technology. 2017;**62**:104-112

[12] Lafarga T, Queralt AV, Bobo G, Abadias M, Aguiló-Aguayo I. Thermal processing technologies. In: Pathania S, Tiwari BK, editors. Food Formulation: Novel Ingredients and Processing Techniques. John Wiley & Sons, Ltd; 2021. pp. 165-181

[13] Marra F, Zhang L, Lyng JG. Radio frequency treatment of foods: Review of recent advances. Journal of Food Engineering. 2009;**91**:497-508

[14] Gavahian M, Tiwari BK, Chu Y-H, Ting Y, Farahnaky A. Food texture as affected by ohmic heating: Mechanisms involved, recent findings, benefits, and limitations. Trends in Food Science and Technology. 2019;**86**:328-339

[15] Yildiz-Turp G, Sengun IY, Kendirci P, Icier F. Effect of ohmic treatment on quality characteristic of meat: A review. Meat Science. 2013;**93**:441-448

[16] Varghese KS, Pandey MC, Radhakrishna K, Bawa AS. Technology,

applications and modelling of ohmic heating: A review. Journal of Food Science and Technology. 2014;**51**:2304-2317

[17] Engchuan W, Jittanit W, Garnjanagoonchorn W. The ohmic heating of meat ball: Modeling and quality determination. Innovative Food Science and Emerging Technologies. 2014;**23**:121-130

[18] Park I-K, Kang D-H. Effect of electropermeabilization by ohmic heating for inactivation of Escherichia coli O157: H7, salmonella enterica serovar typhimurium, and listeria monocytogenes in buffered peptone water and apple juice. Applied and Environmental Microbiology. 2013;**79**:7122-7129

[19] Ito R, Fukuoka M, Hamada-Sato N. Innovative food processing technology using ohmic heating and aseptic packaging for meat. Meat Science. 2014;**96**:675-681

[20] Somavat R, Mohamed HMH, Chung Y-K, Yousef AE, Sastry SK. Accelerated inactivation of Geobacillus stearothermophilus spores by ohmic heating. Journal of Food Engineering. 2012;**108**:69-76

[21] Somavat R, Mohamed HMH, Sastry SK. Inactivation kinetics of Bacillus coagulans spores under ohmic and conventional heating, LWT-food. Science and Technology. 2013;**54**:194-198

[22] Piyasena P, Dussault C, Koutchma T, Ramaswamy HS, Awuah GB. Radio frequency heating of foods: Principles, applications and related properties—A review. Critical Reviews in Food Science and Nutrition. 2003;**43**:587-606

[23] Rowley AT. Radio frequency heating. In: Richardson PS, editor. Thermal technologies in food processing. Cambridge, U.K.: Woodhead Publishing; 2001:163-177

[24] Ohlsson T. Minimal processing of foods with electric heating methods. In: Oliveira FAR, Oliveira JC, Hendrickx ME, Knorr D, Gorris L, editors. Processing foods: Quality optimisation and process assessment. Boca Raton, FL: CRC Press; 1999. pp. 97-105

[25] Li P, Sun Z, Ma M, Jin Y, Sheng L. Effect of microwave-assisted phosphorylation modification on the structural and foaming properties of egg white powder. Lwt. 2018;**97**:151-156

[26] Kozempel MF, Annous BA, Cook RD, Scullen OJ, Whiting RC. Inactivation of microorganisms with microwaves at reduced temperaturas. Journal of Food Protection. 1998;**61**:582-585

[27] Thirumdas R, Sarangapani C, Annapure US. Cold plasma: A novel non-thermal technology for food processing. Food Biophysics. 2015;**10**:1-11

[28] Dolatowski ZJ, Stadnik J, Stasiak D. Applications of ultrasound in food technology. Acta Scientiarum Polonorum. Technologia Alimentaria. 2007;**6**:88-99

[29] Mason TJ, Chemat F, Ashokkumar M. Power ultrasonics for food processing. In: Ashokkumar M, editor. Power Ultrasonics: Applications of High-Intensity Ultrasound. Cambridge: Elsevier Ltd; 2015. pp. 815-843

[30] Bhargava N, Mor RS, Kumar K, Sharanagat VS. Advances in application of ultrasound in food processing: A review. Ultrasonics Sonochemistry. 2021;**70**:105293

[31] Rendueles E, Omer MK, Alvseike O, Alonso-Calleja C, Capita R, Prieto M. Microbiological food safety assessment

of high hydrostatic pressure processing: A review, LWT-food. Science and Technology. 2011;**44**:1251-1260

[32] Considine KM, Kelly AL, Fitzgerald GF, Hill C, Sleator RD. High-pressure processing–effects on microbial food safety and food quality. FEMS Microbiology Letters. 2008;**281**:1-9

[33] Huang H-W, Hsu C-P, Wang C-Y. Healthy expectations of high hydrostatic pressure treatment in food processing industry. Journal of Food and Drug Analysis. 2020;**28**:1-13

[34] Gaougaou G, Shankar S, Liot Q, Constant P, Déziel E, Lacroix M. Gamma irradiation triggers a global stress response in Escherichia coli O157: H7 including base and nucleotides excision repair pathways. Microbial Pathogenesis. 2020;**149**:104342

[35] Wang S, Zhang Y, An W, Wei Y, Liu N, Chen Y, et al. Magnetic relaxation switch immunosensor for the rapid detection of the foodborne pathogen salmonella enterica in milk samples. Food Control. 2015;**55**:43-48

[36] Vorobiev E, Lebovka N. Pulsed electric field in green processing and preservation of food products. In: Chemat F, Vorobiev E, editors. Green Food Processing Techniques. France: Elsevier Inc.; 2019. pp. 403-430

[37] Huang K, Tian H, Gai L, Wang J. A review of kinetic models for inactivating microorganisms and enzymes by pulsed electric field processing. Journal of Food Engineering. 2012;**111**:191-207

[38] Preetha P, Pandiselvam R, Varadharaju N, Kennedy ZJ, Balakrishnan M, Kothakota A. Effect of pulsed light treatment on inactivation kinetics of Escherichia coli (MTCC 433) in fruit juices. Food Control. 2021;**121**:107547

[39] Timmermans RAH, Mastwijk HC, Berendsen L, Nederhoff AL, Matser AM, Van Boekel M, et al. Moderate intensity pulsed electric fields (PEF) as alternative mild preservation technology for fruit juice. International Journal of Food Microbiology. 2019;**298**:63-73

[40] Tiwari BK, Brennan CS, Curran T, Gallagher E, Cullen PJ, O'Donnell CP. Application of ozone in grain processing. Journal of Cereal Science. 2010;**51**:248-255

[41] Mohammad Z, Kalbasi-Ashtari A, Riskowski G, Juneja V, Castillo A. Inactivation of salmonella and Shiga toxin-producing Escherichia coli (STEC) from the surface of alfalfa seeds and sprouts by combined antimicrobial treatments using ozone and electrolyzed water. Food Research International. 2020;**136**:109488

[42] Giménez B, Graiver N, Giannuzzi L, Zaritzky N. Treatment of beef with gaseous ozone: Physicochemical aspects and antimicrobial effects on heterotrophic microflora and listeria monocytogenes. Food Control. 2021;**121**:107602

[43] Ekezie F-GC, Sun D-W, Cheng J-H. A review on recent advances in cold plasma technology for the food industry: Current applications and future trends. Trends in Food Science and Technology. 2017;**69**:46-58

[44] Lin L, Liao X, Li C, Abdel-Samie MA, Cui H. Inhibitory effect of cold nitrogen plasma on salmonella typhimurium biofilm and its application on poultry egg preservation. LWT. 2020;**126**:109340

[45] Devi Y, Thirumdas R, Sarangapani C, Deshmukh RR, Annapure US. Influence of cold plasma on fungal growth and aflatoxins production on groundnuts. Food Control. 2017;**77**:187-191

Section 2

Food Processing and Food Products

Chapter 2

Chemistry of Camel Milk Proteins in Food Processing

Roua Lajnaf, Hamadi Attia and Mohamed Ali Ayadi

Abstract

Camel milk and its extracted protein fractions were found to provide various potential techno-functional properties which can be used in the food industry. This chapter summarizes existing knowledge on camel milk protein's chemistry to explain the different reactions and their control for the major processes utilized by the modern milk processing industry. The composition and chemical properties of camel milk proteins including caseins and whey proteins are investigated. The effect of processing upon denaturation, aggregation, and destabilization of milk proteins is updated. Technological consequences of thermal processing as well as techno-functional properties of camel milk proteins are also described in different techno-functional properties including foaming, emulsifying, and gelling properties. This chapter aims to improve camel milk production and consumption worldwide not only in the arid countries and the hot regions.

Keywords: camel milk, food industry, caseins whey proteins, food processing, thermal processing

1. Introduction

According to recent statistics by the statistics of Food and Agriculture Organization [1], the total population of camels in the world is estimated to be about 38.6 million, with Chad having the largest herd worldwide (8.8 million) followed by Somalia (7.3 million), Sudan (4.9 million), and Kenya (4.7 million) [1]. Camels live mainly in the vast pastoral areas in Asia and Africa, they are divided into two different species belonging to the genus Camelus. Dromedary camels (*Camelus dromedaries*) with one-humped and Bactrian camel (*Camelus bactrianus*) with two-humped [2]. Overall, dromedary camels mainly live in desert arid areas including the Middle East, North and East Africa, South West Asia and Australia while Bactrian camels prefer living in cooler areas such as East to Northern China, West Asia, and Southern Russia (Mongolia and Kazakhstan) [2]. Camels are usually considered to be a good source of milk and meat, meanwhile they are used for other purposes such as sports racing and transportation [3].

Camel milk plays a key role in human nutrition, especially in hot regions and arid countries. Indeed, this milk contains all the essential nutrients already found in bovine milk [4, 5]. According to the latest FAO statistics, camel milk production (both species) in the world is reported to be about 3.11 million tons per year representing

 IntechOpen

0.34% of the total milk production of the world, whereas the cow milk production represents 81.2% of total milk production (746 million tons per year) [6].

In Tunisia, total camel milk production is estimated to be around 1099.64 tons per year, representing only 0.1% of total milk production in Tunisia [1, 7]. Camel milk is popular in Tunisia and consumed as fresh milk as a treatment for a series of diseases such as cancer diseases. The produced camel milk in Tunisia is also dedicated to scientific research in various laboratories and research centers. Indeed, recently, camel milk was also reported to be an efficient treatment for other diseases, such as dropsy, tuberculosis, jaundice, hepatitis, asthma, and leishmaniasis [8]. Camel milk has also other potential therapeutic properties, such as anti-carcinogenic, anti-diabetic, and anti-hypertensive and has been recommended to be consumed by children who are allergic to bovine milk [9–14].

Unfortunately, camel milk has not been given as much attention in research compared with cow milk because of its relatively limited production and consumption despite its health benefits and therapeutic properties. Most of the research conducted on camels in the past was mainly focused on their physicochemical features. However, recent studies have mainly concentrated on the compositional, characteristics and technological properties of camel milk and its derived proteins. This review covers the recent works on camel milk properties with an emphasis on camel milk proteins. The aim of this chapter is to review the currently available information on Dromedary camel milk properties, composition as well as camel milk proteins: extraction processes, biochemical, and techno-functional properties.

2. Protein composition of camel milk

Overall, milk proteins represent a significant nutritional intake (source of essential amino acids). These proteins represent also a source of important techno-functional properties for the conservation and processing of milk into dairy products for human consumption [15].

The total protein content in camel milk ranges from 21.5 to 49 g/L with an average of 31 g/L of milk [16]. This variation in the composition of camel milk proteins depends not only on the race of the producing female but also on seasonal conditions [17]. For instance, protein contents in camel milk, which was collected from the same breed, were found to vary significantly depending on seasons ranging between 24.8 g/L of proteins in summer to 29 g/L in winter [18].

As with other milk of different mammalian species, dairy proteins are commonly classified according to their solubility in two fractions: caseins (insoluble in acidic medium) and whey proteins (called soluble proteins) (**Table 1**). Indeed, the caseins precipitate at their isoelectric pH which is 4.6 and 4.3 for bovine and camel milk, respectively, while whey proteins remain soluble in these pH values [19–22].

2.1 Caseins

Camel caseins are phosphoproteins that represent the most abundant protein fraction of milk. They occupy 61.8–88.5% of all camel proteins with an average of 75.4% (w/w) against an average content of 80% (w/w) for cow's milk [23].

Compositionally, caseins in bovine milk are composed of four caseins including α_{S1}-, α_{S2}-, β-, and κ-caseins with a molar ratio of approximately 4:1:4:1 in bovine milk [22].

	Cow's milk		Camel's milk	
	Concentration (g/L)	**MW**	**Concentration (g/L)**	**MW**
Fat globules	40	0.1–15 μm (average = 3.78 μm)	12–64 (average = 35)	2.99 μm
Caseins micelles	26	100–140 nm (average = 120 nm)	16.3–27.6	260–300 nm (average = 280 nm)
Whey proteins	7	3–6 nm	6.3–8	n.d
α-La	1.2	14 kDa	>5	14.430 kDa
β-Lg	3.2	18 kDa	—	—
SA	0.4	66 kDa	3.4	66 kDa
Ig	0.8	150–900 kDa	0.718	80 kDa
Lf	0.1	86 kDa	0.229	75 kDa
Lactose	46	0.35 kDa	24–58 (average = 44)	0.35 kDa
Minerals	7	—	6–9 (average = 7.9)	—
Vitamins	3.2	—	3.7	—

Abbreviations: α-La: α-lactalbumin; β-Lg: β-lactoglobulin; SA: serum albumin; Ig: immunoglobulins; Lf: lactoferrin [15].

Table 1.
Composition of camel milk in comparison with cow's milk.

On the other hand, camel caseins consist of the known four sub-fractions including α_{S1}-, α_{S2}-, β-, and κ-caseins with proportions approximately being 22, 9.5, 65, and 3.5%, respectively in bovine milk (**Figure 1**) [25]. Recently, Lajnaf et al. [26] found that camel sodium caseinates contain four caseins at different percentages 1.1, 45.5, and 53.4% for κ-, α-, and β-caseins, respectively. The caseins of camel milk are homologous to bovine caseins with identity levels that range between 44.6% (α_{S1}-casein) and 67.2% (β-casein) [27]. The α- and β-caseins are known as calcium-sensitive caseins or "sensitive calcium caseins" due to their precipitation at a calcium concentration estimated at 30 mM, while κ-casein remains in solution under these conditions.

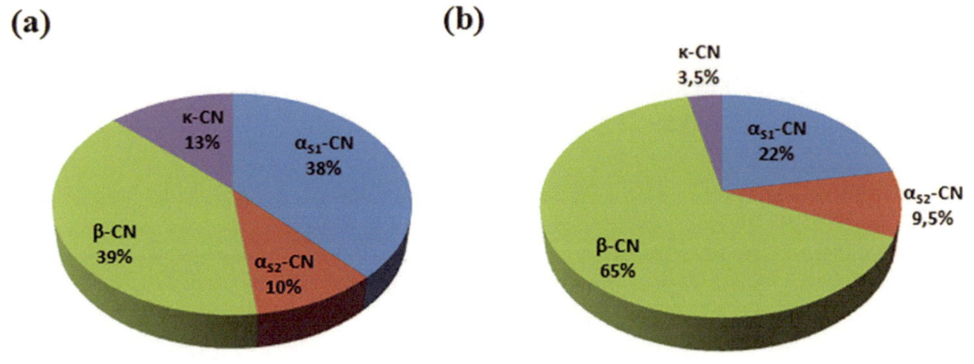

Figure 1.
Proportions of the different caseins α_{S1}-, α_{S2}-, β and κ of the total caseins of cow's milk (a) and camel's milk (b) [24]. Abbreviation: CN: casein.

Camel milk is distinguished by the low contents of κ-casein as reported by various authors. In the same way, Lajnaf et al. [28] found that no peaks were detected for the κ-casein due to its low concentration which probably makes it obscured by other caseins.

Overall, the comparison of camel and bovine properties revealed that camel milk caseins are less phosphorylated than their bovine counterparts and less negatively charged at neutral pH when compared to bovine caprine caseins [29, 30].

2.1.1 α-Casein

The α-casein, which includes both α_{S1}- and α_{S2}-caseins, is the most abundant protein in cow's milk and its concentration in milk is estimated at 12.8 ± 2.3 g/L. However, the concentration of this protein is lower in camel milk (7.6 g/L) [25, 31, 32].

The α_{S1}-casein, whose concentration is round of 9.5 and 5.3 g/L in bovine and camel milk, respectively, representing 38 and 22% of total bovine and camel caseins, respectively. Bovine α_{S1}-casein contains 199 amino acid residues with a molecular weight (MW) estimated at 22.9 kDa, while camel α_{S1}-casein is slightly bigger with 215 amino acids and a MW of 25.8 kDa. The isoelectric point (pI) is estimated to be around 4.26 and 4.40 for the bovine and camel α_{S1}-caseins, respectively [25, 33]. The differences between camel and bovine α_{S1}-caseins result in identity and similarity indexes which are around 44.6 and 59.7%, respectively [27]. Bovine α_{S1}-casein is characterized by the absence of cysteine residues. However, it contains 8 serine residues in phosphorylated form. Due to the presence of a large number of proline residues (9.2 and 8.5% proline respectively for camel and bovine α_{S1}-caseins, respectively) [33].

The content of α_{S2}-casein in camel milk is similar to that of cow's milk. It represents 10 and 9.5% of the caseins of bovine and camel caseins, respectively [34]. Recently, this content has been reported as 0.3–3.9 g/L as reported by Mohamed et al. [35]. The primary structure of α_{S2}-bovine casein has 207 amino acid residues with an MW of 24.4 kDa, while camel α_{S2}-casein has a lower MW of 22 kDa as it contains 178 residues of amino acids. The pI is estimated respectively at 4.78 and 4.58 for both bovine and camel α_{S2}-caseins [33].

It is well known that the α_{S2}-casein is the most hydrophilic of the other caseins. It has 11 residues of phosphorylated serines and is characterized by the presence of two cysteine residues (residues 36 and 40) forming intramolecular disulfide bridges. This casein is found in partly milk in dimeric form, the two polypeptide chains of which are connected by two disulfide bridges [25]. Its secondary structure contains 32% α-helix and 30% β-sheets leading to a more organized and structured conformation when compared to those of α_{S1}-casein. Similarly to other milk proteins, the differences between camel and bovine α_{S2}-caseins result in identity and similarity indexes which are around 58.3 and 69.2%, respectively [27].

2.1.2 β-Casein

The β-casein is the main protein in camel milk with a concentration that ranges between 12.8 and 15 g/L representing 65% of the total caseins of camel milk according to Kappeler et al. [34]. However, recent works noted a minimization of its proportion to 53.4% [26], 44.8%, and even 30% according to Felfoul et al. [36] and Ereifej et al. [23], respectively.

The β-caseins of cow's and camel's milk showed differences in their structures and physicochemical characteristics. In fact, bovine β-casein is composed of 209 amino acid residues with an MW of 23.5 kDa and a pI of 4.49, while camel β-casein is slightly bigger than its bovine counterpart as it contains 217 amino acids leading to an MW of 24.9 kDa and a pI of 4.66. The rates of similarity (84.5%) and identity (67.2%) of the β-caseins are higher than those found for other caseins as reported by Lajnaf et al. [27] and Barzegar et al. [37].

The β-casein, the most hydrophobic of all the caseins, is characterized by a very high amphipolar character. Indeed, it has a C-terminal part (residues 136–209) which is very rich in hydrophobic amino acids, while its N-terminal part is hydrophilic and contains phosphorylated residues (residues 1–40) providing additional negative charges to the molecule. This protein is also characterized by the absence of disulfide bridges, which gives it significant resistance to heat treatment. β-Casein is classified as an intrinsically unstructured protein thanks to the large number of proline residues (16.7%) preventing the formation of secondary structures. Due to its particular structure (unordered, high hydrophobicity, relatively low molecular mass and absence of disulfide bridges), this casein is often at the origin of the properties sought in "stabilizing" protein food ingredients used in the dairy industry [38].

2.1.3 κ-Casein

The κ-casein is the key milk protein that is involved in the rennet coagulation process of milk. Among caseins, the concentration of κ-casein (4.4 ± 0.3 g/L) was found the lowest representing 13% of bovine caseins. However, it is found in camel milk at a content four times lower than that of cow's milk varying from 0.1 to 2.4 g/L representing 3.5% of caseins or even 1.1% [26, 29, 35]. Bovine κ-casein has 169 amino acid residues with an MW of 18.9 kDa and a pI of 3.97, while camel κ-casein is composed of 162 amino acids with a molecular mass of 18.2 kDa and a pI of 4.11 [33]. These differences between both camel and bovine κ-caseins result in similarity and identity of 58.4 and 66.3%, respectively.

Similarly to β-casein, κ-casein has a particular amphipolar structure with a C-terminal part that contains highly hydrophilic residues and a hydrophobic N-terminal part. It is also characterized by a low calcium binding capacity due to the presence of a single phosphorylation site at position 149.

It is well known that the partial hydrolysis of bovine κ-casein by chymosin takes place at the peptide bond 105(Phe)-106(Met) leading to the release of a very hydrophilic peptide: the caseinomacropeptide (64 amino acids—molecular mass 6.7 kDa) and the formation of paracasein κ, which is very hydrophobic and insoluble. The cleavage site of camel κ-casein by chymosin is located at position 97(Phe)-98(Ile) (**Figure 2**) and leads to the release of a macropeptide with a molecular mass of 6, 77 kDa which is comparable to bovine macropeptide [29].

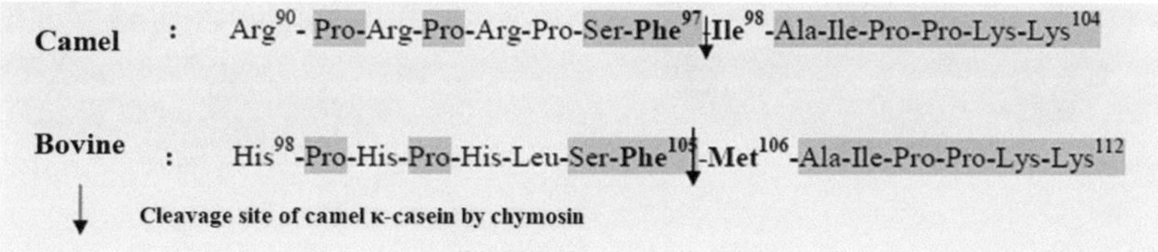

Figure 2.
Cleavage sites of camel and bovine κ-caseins by chymosin [15].

2.2 Whey proteins

Whey proteins represent the second protein milk fraction representing 20–25% (w/w) of total milk proteins depending on the milk origin [39]. Camel whey proteins accounted on average for 24.51% of the total protein ranging between 11.49 and 38.82% of total milk proteins [23]. Overall, extracted camel whey after acid precipitation of caseins at pH 4.3 has a white color compared to the yellowish color of bovine whey. This is due to the low content of riboflavine in camel whey [2].

Generally, the protein composition of whey varies according to the mammalian specie. For instance, the soluble fraction of cow's milk, the protein composition is thoroughly studied: β-lactoglobulin is the main protein (~55%), followed by α-lactalbumin (~25%), the albumin serum (SA) (15%), and finally the immunoglobulins (5%) (**Table 1**). Camel whey is distinguished by the total absence of β-lactoglobulin similar to human milk [28, 33, 34, 40]. Thus, α-lactalbumin is the major protein of camel whey 50–54% of all of the globular proteins in this milk, this protein is followed by camel serum albumin (CSA) (36%), lactoferrine (2%), and immunoglobulins (8%) (**Figure 3**) [33].

Several works have shown that camel whey contained other specific protein components such as the PGRP (Peptidoglycan Recognition Protein), lactophorine, Wap (Whey Acidic Protein), and CWBP (Camel Whey Basic Protein) [29, 33, 41].

2.2.1 α-Lactalbumin

The α-lactalbumin (α-La) is the major protein in camel whey as the β-lactoglobulin which is the major protein in bovine whey is totally absent [27, 28, 36, 40, 43]. The concentration of this protein in camel milk is significantly higher than that of cow's milk (1.08 g/L) [32] as it ranges between 2.1 g/L according to Omar et al. [32] and 5 g/L according to El-Agamy [33].

The primary sequence of camel α-La was determined by Beg et al. [42]. As its bovine counterpart, camel α-La is composed of 123 amino acids, in which 39 residues are different when compared to bovine α-La. Consequently, the similarity and identity levels between these proteins according to the sequence alignment data are

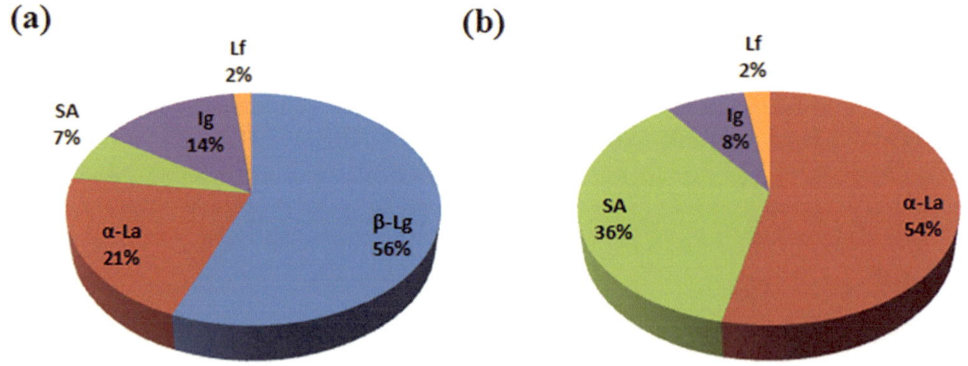

Figure 3.
Proportions of the different whey proteins of cow's milk (a) and camel's milk (b) according to El-Agamy [33]. Abbreviations: β-Lg: β-lactoglobulin, α-La: α-lactalbumin, SA: serum albumin, Ig: immunoglobulins, Lf: lactoferrin.

82.9 and 69.1%, respectively according to Salami et al. [24]. MW and pI of camel α-La (MW = 14.43 kDa and pI = 4.87) are slightly higher than those of bovine α-La [43, 44].

Similarly to its bovine counterpart, camel α-La has a high affinity for the Ca^{2+} ion with a higher exposure of hydrophobic groups upon calcium depletion than the bovine α-La [45, 46]. In terms of nutritional properties, several studies have shown that camel α-La is characterized by a higher digestibility than that of bovine milk, as well as greater antioxidant activity with respect to Ferric-reducing antioxidant power, iron chelating, and antiradical activities especially in their apo forms [44]. This protein presented in its apo form great antibacterial and antifungal properties toward various pathogenic species [43, 44].

2.2.2 Camel serum albumin

Serum albumin (SA) protein is a whey protein characterized by its relatively high MW. Indeed, bovine serum albumin (BSA) consists of 583 amino acids with an MW of 66.4 kDa, its primary sequence was determined Hirayama et al. [47]. It has 17 intramolecular disulfide bridges and a free thiol group. On the other hand, camel serum albumin (CSA) was identified by SDS-PAGE as a similar protein to its bovine counterpart with the same MW (66 kDa) [15, 29].

BSA and CSA were reported to have similar concentrations ~0.4 g/L with different proportions among whey proteins (1.5 and 7% of total bovine and camel whey proteins fractions, respectively). However, the contents of CSA are higher in camel colostrum with concentrations greater than 3.4 g/L [48].

2.2.3 Minor camel whey proteins

Lactoferrin is a glycoprotein that belongs to the transferrin family. It contains two binding sites for iron cations and more preferentially the ferric ion (Fe^{3+}). This ability to scavenge iron persists even at low pH values in the stomach and intestines, to deplete free iron which could slow down bacterial growth in the intestines [29]. The concentration of lactoferrin in milk varies according to the producing animal species and according to the stage of lactation. Camel milk is very rich in lactoferrin compared to the milk of other mammalian species. This richness is a form of adaptation to difficult living conditions for young camels to make them more resistant to infections [49].

Camel lactoferrin is composed of 689 amino acids with an MW of 75.3 kDa. The primary sequence of camel lactoferrin has a similarity level of 91.6% with its bovine and human counterparts and 91.3% with porcine lactoferrin. It is a basic protein with a pI of around 8.14 (compared to a value of 8.18 for bovine lactoferrin) [33].

PGRP or "Peptidoglycan Recognition Protein" is part of a family of proteins described recently. It is known for its action on gram-positive bacteria as well as other microorganisms such as nematodes. This inactivation of pathogens is carried out by the binding of this protein to the peptidoglycan of the bacterial membrane, hence its name "Peptidoglycan recognition protein" or PGRP [50]. PGRP is a protein that is not detected in cow's milk. It was isolated from camel milk by Kappeller et al. [50]. It is a protein which is characterized by its low molecular mass (19.11 kDa) containing 172 amino acids. The PGRP of camel milk is a basic protein, it is very rich in Arg residues whereas it is poor in Lys. It is found in camel milk at a concentration of 1.74 g/L [32]. The pI of camel PGRP is 8.73 which is higher than that of human PGRP (pI = 7.94). The similarity level between PGRP in camel milk and human milk is around 91.2% [33]. The PGRP content increases in camel milk in case of infection of the mammary glands. Also, the high level of PGRP in camel milk at the start of lactation contributes to the protection of the mammary gland as well as the transmission of immunity to the newborn [50].

Camel Whey Basic Protein or CWBP (Camel Whey Basic Protein) is also a protein specific to camel milk. It was identified from camel whey by SDS-PAGE electrophoresis [51] and by ion exchange chromatography [48]. This protein, of relatively low MW (20 kDa), has a unique structure and has no analogy with other dairy proteins. It has been demonstrated in the whey of camelids of the dromedary and bacterial species.

WAP or Whey Acidic Protein is a soluble protein found in the milk of certain mammalian species including rabbits, pigs, rodents, camelids and humans. WAP is a whey protein found at a concentration of 0.157 g/L in camel milk. It contains 117 amino acids with an MW of 12.56 kDa. WAP consists of two domains with four disulfide bridges with a pI of 4.5 [33]. Thus, camel milk contains the highest rate of natural bioactive components, which explains its long shelf life compared to cow's milk [33].

3. Effect of processing on chemistry of camel milk proteins

Thermal treatments are important food processes including in most dairy industries to obtain bacteriologically safe final products and to extend their shelf life. However, a number of structural modifications have been reported and noted in the milk protein components depending on temperature time, and rate of heating. For instance, Singh [52] reported that a range of large heterogeneous protein aggregates of milk proteins occurred in heat-treated milk. Indeed, the association of heat-induced milk proteins which are occurring under different heating conditions has been extensively studied by various authors [53].

Overall, both caseins and whey proteins in heat-treated milk are engaged in protein denaturation. Furthermore, the formation of intermolecular disulfide bridges is mostly responsible for heat-induced protein association in milk. Thermal protein denaturation has been acknowledged as the first step of the reactions leading to the aggregation of the disulfide-linked milk proteins. The resulted thiol groups of cysteine residues which are appearing in unfolded proteins, can initiate thiol-disulfide exchange reactions within hydrophobically-linked protein aggregates. On the other hand, self-aggregation of heat-denatured β-lactoglobulin in cow's milk, and heat-induced association of various whey proteins and their aggregates with caseins have been investigated and explained according to this mechanism [54].

3.1 Effect of processing on caseins

Similarly to cow's milk, camel milk proteins are significantly affected by thermal treatment processing. However, only few studies about the effect of heat treatments on camel milk proteins including caseins and whey proteins are available in the literature [55].

First, Felfoul et al. [36] found using LC-MS and SDS PAGE electrophoreses techniques that after heating camel milk at 80°C for 60 min, various significant modifications in protein composition were observed.

Indeed, these authors noted that fresh camel milk contains α-La, PGRP, CSA, and caseins proteins as major proteins. In the same way as bovine milk, the thermal treatment of camel milk at 80°C for 60 min caused various significant modifications in proteins including whey proteins and caseins. However, camel α_{S2}-, β-, and γ-caseins concentrations have not been significantly modified by heat treatment similarly to bovine caseins. Other study revealed that the effect of the heating temperature increases on camel milk was mild on β-casein and both αS1- and αS2-caseins, whereas it was drastic on κ-casein. Indeed, electrophoretic bands of whey proteins including CSA and α-La as well as κ-casein decreased at 90°C [56].

On the other hand, Lajnaf et al. [26] investigated the effect of different heating temperatures on extracted camel sodium caseinates at neutral pH. RP-HPLC results of these authors showed that both bovine and camel caseins peaks including κ-casein, α-casein and β-casein remained almost intact upon heating at 70 and 80°C for 30 min. However, higher temperatures (90 and 100°C) significantly affected camel casein peaks especially α-casein and β-casein, which decreased significantly at these temperatures. Furthermore, the degradation of caseins is synchronized by the appearance of new protein fractions after heating at 90°C for 30 min. In the same way, new peptides were generated upon heating from the parent caseins. Thus, the heat treatment of camel caseinates solutions results in the degradation polymerization of proteins as well as the liberation of several peptides due to protein degradation [26].

3.2 Effect of processing on whey proteins

The effect of processing on camel whey proteins especially thermal processing as well as the acidification process is being studied by many researchers in recent scientific works who are interested in the valorization of camel milk and its consumption as a new alternative of bovine whey especially due to the total absence of the β-lactoglobulin in camel milk.

First, the work of Felfoul et al. [21] was considered as the first study leading to understanding the chemistry of camel whey proteins upon heating and at different pH levels as they studied the effect of different heating temperatures on sweet and acid camel whey. These authors noted that protein denaturation started after heating whey for 30 min for all temperatures. The whole phenomenon happened during 30 min of heating. The obtained results by these authors have shown that heating both bovine and camel whey at 60°C does not generate any denaturation phenomena as it is already observed by Laleye et al. [40]. The electrophoresis patterns showed also that heating camel whey at 90°C during 30 min CSA band disappearance for both rennet and acid wheys. On the other hand, α-La concentration decreased as a function of heating temperature.

As previously reported, the major camel whey proteins are α-La, CSA, and PGRP [36, 41, 48, 50, 57]. These proteins were significantly affected by heat treatment at 80°C for 60 min as revealed by Felfoul et al. [36]. Indeed, the corresponding peak of CSA decreased significantly after heating at this temperature while camel α-La and PGRP have completely disappeared from the HPLC-UV chromatograms. Indeed, these authors found that the concentration of CSA in fresh camel milk was decreased by 42%, while PGRP concentration decreased by 68%, whereas, there was 100% of α-La disappeared from camel milk. Thus, the most heat-sensitive whey protein in camel milk obviously corresponds to camel α-La followed by PGRP and CSA [21, 36]. In the same way, Lajnaf et al. [28] found that the chromatographic peak of the α-La began to decline after the heat treatment at 70°C for 30 min, it decreased significantly when the heating temperature raises from 80 to 100°C for 30 min. Thus, the reduction of the chromatogram peaks is the consequence of the protein denaturation and aggregation upon heating [28]. However, for bovine milk, the peaks of the α-La and the β-Lg started immediately diminished after the heat treatment at 80°C for 30 min.

The peak of β-Lg totally disappeared after heating at 90 and 100°C for 30 min, unlike the β-Lg dimer peak that increased due to the creation of heat-induced disulfide-bonded dimers as intermediates in the whey proteins aggregation [28].

On the other hand, differential scanning calorimetry (DSC) thermograms of Felfoul et al. [21] showed that denaturation temperatures of camel α-La were 73.8°C in camel rennet whey and 60.5°C for camel acid whey. Atri et al. [45] noted that denaturation temperatures of purified camel α-La are 71.7 and 39.6°C in its holo (calcium loaded) and apo (calcium depleted) forms. Indeed, the absence of β-lactoglobulin in camel milk whose denaturation temperatures are 79.6 and 83.4°C in sweet and acid bovine wheys, respectively resulted from different denaturation and aggregation phenomena during heat treatment [21].

Other scientific works have shown that the combination of heating treatment and acidification of camel wheys induced an immediate disappearance of the α-La and the appearance of several intermediate protein species including dimers, trimers of α-La. These protein species were formed during heating and before aggregation [20]. These authors have found that acid wheys carried higher denaturation levels compared to sweet wheys regardless of heating temperature value. These findings confirmed that acid whey is characterized by a higher thermal sensitivity than the sweet one with the higher thermal sensitivity of camel whey proteins compared to bovine whey proteins especially at neutral conditions [20]. In the same way, Laleye et al. [40] noted that camel milk whey proteins are slightly more susceptible to heat denaturation than bovine whey proteins regardless of pH level. This behavior can be explained by the particular structure of camel α-La, especially in acidic conditions. Lajnaf et al. [41] reported that the open structure of the camel α-La molecule and the reduced electrostatic repulsion of this protein near its pI are all factors that could promote the creation of large aggregates. In the same way, Lajnaf et al. [57] observed that the purified camel α-La isolated from camel milk was more flexible in acidic conditions, regardless of heating temperature, due to the reduced negative charge of this protein and its molten globular state at low pH values.

Recently, Lajnaf et al. [43] reported that there are various structural differences between the camel and bovine α-La as a function of different denaturing conditions in food processing including pH, heating temperature, and guanidine hydrochloride mediated. Camel α-La showed higher stability toward thermal treatments and pH-mediated denaturation. However, it was less stable toward guanidine-mediated denaturation with a fast aggregation and a more disordered structure when compared to its bovine counterpart [43, 58].

4. Effect of processing on camel milk protein functionality

4.1 Foaming properties

The foaming and stabilizing properties of camel milk as well as its protein fractions were investigated by different authors [28, 41, 59–61]. First, Lajnaf et al. [26, 28] studied the effect of different heating temperatures ranging between 70 and 100°C on skimmed camel milk as well as extracted sodium caseinates. These authors noted that for the camel milk and sodium caseinates, heating improved significantly the foamability in comparison with that of bovine milk and bovine caseinates, with better foaming capacity achieved after a heat treatment at 90 and 100°C due to the presence of higher amounts of β-casein in camel milk. Indeed, this protein is well known as a mobile disordered protein due to its particular flexible structure [62]. However, lower foam stability of camel milk and camel caseinates foams is observed due to the different protein composition of both milk proteins especially the absence of β-Lg and the lower amounts of κ-casein [26, 28]. On the other hand, the stability of foam formed from skimmed camel and bovine milk increased significantly with increasing preheating temperatures up to 90°C, above which lower foam stability is observed [28]. While for camel sodium caseinates, foam stability increased as a function of heating temperature even at 100°C [26]. The increase in foaming properties of camel milk is attributed to an increase in the hydrophobic interactions due to an exposure of hydrophobic groups, which are already buried inside the globular structure of whey proteins [28, 63]. Furthermore, this behavior can be explained by the increase in the adsorption velocity and the diffusion of milk proteins upon heating at the air-water interface as confirmed by Dickinson [64]. In the same way, heat treatment significantly ameliorated the foaming properties of camel and bovine sodium caseinates especially at high-temperature values (90 and 100°C for 30 min).

Parallely, the heating process affects the physicochemical properties of caseins including the increase of surface hydrophobicity due to the greater exposure of buried hydrophobic groups and the increase of the ability to reduce the interfacial tension at the air-water interface. The heating process also decreases the electronegative charge of proteins leading to a greater flexibility and hence, higher foaming properties of heated milk proteins [26].

The foaming properties of camel whey proteins are significantly affected by different stabilizing food processing, especially thermal processing and acidification. This behavior is mainly explained by scientists as camel milk is totally deficient in β-Lg because it is well-known that this protein plays a key role in the process of protein aggregation in bovine whey solution [60].

Camel whey foaming properties are reported to depend on both pH value and thermal treatments. Camel whey solutions showed the best foamability closed off the isoelectric point of camel α-La (around pH 4.3), regardless of heating temperature. Thermal treatments at 70°C significantly improved the foaming properties of both bovine and camel acid wheys. However, the stability of foam greatly increased upon heating only for the acid camel whey [41]. Acid camel whey is distinguished by its exceptional ability to create foams with the greatest foam stability if compared to other whey, with an increase of these properties after a heat treatment. Hence, the lack of β-Lg in camel whey leads to exceptional foaming properties of this whey, especially with the combination of preheating and preacidification before the creation of the foams [41]. In the same way, the foamability of camel α-La in solution was maximal in acid conditions, near its effective pI. Indeed, at this acid pH, the protonation of the negative groups decreased the electrostatic repulsions of the α-La and induced a partial denaturation with the release of its chelated calcium. The obtained molten globular state enhanced the foaming properties of this protein. Heating processes improved the stability of the foam which is created by camel α-La due to the presence of aggregated proteins at the air-water interface. Aggregates are reported to contribute to improving foam stability whereas, they slowed the adsorption of proteins and the creation of foam [43, 57]. In addition to the heating process, the effect of the spray drying process on the techno-functional of camel milk proteins was investigated by Zouari et al. [65] noted the low denaturation extent of camel and bovine milk proteins powders participated in the enhancement of their foaming capacity and stability [65].

4.2 Emulsifying properties

Emulsification is a common food process in the food industry, it is encountered with mayonnaise sauces, cream, soups, butter, and margarine [66]. Overall, oil-in-water emulsions are produced by the homogenization process of oil and aqueous phases in the presence of emulsifiers which are adsorbed onto the surfaces of oil-droplets leading to the reduction of the interfacial tension and emulsion creation. In the food industry, the most common emulsifiers used are milk proteins including caseins such as β-casein which is the most surface-active dairy protein, and whey proteins including β-Lg and α-La [62, 66]. The effect of food processing on the ability of camel milk proteins to create and stabilize emulsions was studied by different authors [20, 60, 67]. First, Lajnaf et al. [20] reported that camel whey emulsifying properties depended on both pH level and the degree of denaturation of these proteins after a heat treatment. Higher emulsifying activity stability was obtained for sweet whey especially the sweet camel whey due to the presence of electrostatic repulsive forces between proteins. However, acidification reduces these repulsive forces leading to the reduction of emulsifying properties of milk proteins.

Laleye et al. [40] reported the lower emulsifying properties of pre-acidified camel whey when compared to bovine whey due to the pronounced aggregation of camel whey protein molecules. Indeed, the aggregation behavior of camel whey proteins at lower pH values is associated to the high content of the α-La [40]. Furthermore, thermal treatments of camel whey proteins at 70 and 90°C improved the emulsifying properties of these proteins, especially in acidic conditions due to the denaturation and aggregation of proteins. Indeed, the size of whey proteins' aggregates is higher in acidic conditions than in neutral pH due to the minimized electrostatic repulsion between neigh-boring proteins molecules leading them to interact and aggregate. These aggregates are characterized by a greater ability to stabilize foam and emulsion compared to native proteins [20, 68].

Momen et al. [60] studied the effect of the heating process at 85°C for 15 min in a temperature-controlled water bath on the created emulsions. These authors noted that the emulsions prepared with camel whey proteins did not show any visible aggregation or gelation after heat treatment, whereas emulsions prepared by bovine whey proteins formed a gel-like structure in different protein concentrations. Indeed, the limited heat-induced modification in the conformational structure of camel whey proteins confirmed that these proteins are not very sensitive to heat-induced disulfide bridging and hydrophobic interactions. Thus, this study showed the technological viability of camel whey protein for the fabrication of high-protein emulsion. In the same way, the emulsifying properties of camel α-La were less sensitive to various thermal treatments at 95°C. This behavior was explained by the higher conformational flexibility of this protein which increased with temperature, contrary to its bovine homologous protein [67, 69]. Indeed, bovine α-La enhanced emulsion stability as a function of pH and heat treatment, due to hydrophobic interactions and a more rigid molecular structure compared to camel α-La [69]. Furthermore, a higher surface coverage of the oil droplets was obtained for camel apo α-La which carried the highest ability to reduce the surface tension values at the oil-water interface when compared to bovine α-La in its holo and apo states. The stability of the created emulsions seemed greatest at neutral pH due to the presence of the electrostatic repulsive forces between the adsorbed α-La molecules contrary to these molecules in acidic conditions. These conditions reduced these repulsive forces leading to the decrease of emulsifying properties of camel α-La [43, 44].

Ellouze et al. [70] reported that camel milk β-casein showed an enhanced ability to form softer emulsions and to stabilize oil droplets in acidic conditions, regardless of heat treatment compared to bovine β-casein. However, the heating process affects the interfacial properties of β-casein. Indeed, the ability of this protein to create emulsions did not show any effect upon heating, whereas, stabilization of emulsified oil droplets with β-casein is higher without the heating process as proteins retain their native structure with no thermal denaturation, which allows intramolecular hydrophobic interactions and hence, the maintenance of a stable protein film around the created oil droplets. On the other hand, surface pressure was higher in acidic conditions for camel β-casein and after a thermal treatment at 95°C. This phenomenon is explained by hydrophobic interactions and a relaxed structure allows proteins to be more cohesive under the applied treatments [70]. In the same way, emulsifying properties of camel β-casein solutions depended on pH level with or without thermal processing. Preacidification affects the physicochemical properties of camel β-casein by increasing the surface hydrophobicity and also decreasing the negative charge and the efficiency to reduce the interfacial tension. Therefore, casein precipitation decreases the emulsifying properties of camel β-casein and its ability to create and stabilize emulsions [70].

4.3 Gelling properties

The effect of different food processing on gelling properties of camel milk was described by few scientific works. First, Zouari et al. [71] studied the effect of the acidification process on the gelation of camel milk and found that the gelation behavior of camel milk is mainly controlled by the pI and hydrophobic interactions. These authors found that the intermolecular interactions between different camel milk caseins are higher and stronger when compared with those in bovine milk. The effect of thermal processing on the quality of fermented camel milk products including yogurt and cheese is not completely investigated. The fermentation process of milk into yogurt requires pre-heating in order to denature the whey proteins and form disulfide bridges between these proteins and κ-casein, leading to improved yogurt structure. Manufacturing yogurt from camel milk is difficult and the yogurt curd produced from camel milk is fragile and has a thin consistency because of the presence of bioactive antimicrobial components including PGRP and lactoferrin [2]. Furthermore, the different compositions of camel milk whey proteins, such as its lack of β-lactoglobulin and the predominance of α-lactalbumin are also the reasons for the fragile structure of camel milk yogurt [72]. Pasteurization process of the camel at temperatures higher than 65°C for 30 min results in the manufacturing of camel cheeses with significantly weaker gels [73, 74]. Furthermore, high-pressure processing of milk at 350 MHz for 5 min produces harder cheese than pasteurization treatment at 65°C for 30 min [72]. Finally, further studies are needed to understand and to explain the effect of various food processing, especially thermal ones on gelling properties of camel milk.

5. Conclusion

Camel milk is different in its composition from that of cow's milk including fats, minerals, lactose, and proteins. The main differences in camel milk proteins composition are the total absence of the β-Lg and the low amount of κ-casein, leading to confirm that β-casein and α-La are the major proteins in colloidal and soluble fractions of camel milk, respectively. Camel milk proteins show different behavior when compared to bovine proteins. For instance, previous studies noted that heating treatment of milk significantly affects α-La followed by PGRP and CSA, with a relative thermal sensitivity of whey proteins when compared to caseins. However, thermal treatment of camel caseinates leads to the degradation and denaturation of individual caseins including α-casein and β-casein which is probably associated with liberation of resulted peptides. Different techno-functional properties of camel milk proteins are significantly affected by food processes including thermal processes and nonthermal such as acidification. For instance, the combination of acidification and thermal treatments improves the foaming properties of whey proteins, while these processes reduced emulsifying properties of camel whey proteins.

Finally, this chapter investigates the interesting techno-functional properties and the chemical of camel milk proteins as a function of different food processes. Hence, this could confirm the strong potential of camel milk for potential applications in the food, pharmaceutical, and cosmetic industries.

Author details

Roua Lajnaf[1,2]*, Hamadi Attia[1] and Mohamed Ali Ayadi[3]

1 Alimentary Analysis Unit, National Engineering School of Sfax, Sfax, Tunisia

2 Montpellier University, UMR IATE, Montpellier, France

3 Department of Food Technology, University of Liege—Gembloux Agro-Bio Tech, Gembloux, Belgium

*Address all correspondence to: roua_lajnaf@yahoo.fr; roua.lajnef@enis.tn

References

[1] FAOSTAT, Crops and Livestock Products. 2021. Available from: https://www.fao.org/faostat/en/#data/QCL

[2] Al Kanhal HA. Compositional, technological and nutritional aspects of dromedary camel milk. International Dairy Journal. 2010;**20**:811-821

[3] Osaili TM, Al-Nabulsi AA, Hasan F, Dhanasekaran DK, Hussain AZS, Ismail LC, et al. Effect of eugenol, vanillin and β-resorcylic acid on foodborne pathogen survival in marinated camel meat. Journal of Food Protection. 2023;**86**:100038

[4] Izadi A, Khedmat L, Mojtahedi SY. Nutritional and therapeutic perspectives of camel milk and its protein hydrolysates: A review on versatile biofunctional properties. Journal of Functional Foods. 2019;**60**:103441

[5] Ho TM, Zou Z, Bansal N. Camel milk: A review of its nutritional value, heat stability, and potential food products. Food Research International. 2021;**153**:110870

[6] Lajnaf R, Feki S, Ben Ameur S, Attia H, Kammoun T, Ayadi MA, et al. Cows' milk alternatives for children with cows' milk protein allergy-review of health benefits and risks of allergic reaction. International Dairy Journal. 2023;**141**:105624

[7] Lajnaf R. Camel milk: Nutritional composition , functionality and health benefits—A mini review. Current Trends on Biotechnology & Microbiology. 2020;**2**:389-393

[8] Swelum AA, El-Saadony MT, Abdo M, Ombarak RA, Hussein EOS, Suliman G, et al. Nutritional, antimicrobial and medicinal properties of Camel's milk: A review. Saudi Journal of Biological Sciences. 2021;**28**:3126-3136

[9] El-Agamy EI, Nawar M, Shamsia SM, Awad S, Haenlein GFW. Are camel milk proteins convenient to the nutrition of cow milk allergic children? Small Ruminant Research. 2009;**82**:1-6

[10] Uversky VN, El-Fakharany EM, Abu-Serie MM, Almehdar HA, Redwan EM. Divergent anticancer activity of free and formulated camel milk α-lactalbumin. Cancer Investigation. 2017;**35**:610-623

[11] El-Fakharany EM, El-Baky NA, Linjawi MH, Aljaddawi AA, Saleem TH, Nassar AY, et al. Influence of camel milk on the hepatitis C virus burden of infected patients. Experimental and Therapeutic Medicine. 2017;**13**:1313-1320

[12] Shori AB. Camel milk as a potential therapy for controlling diabetes and its complications: A review of in vivo studies. Journal of Food and Drug Analysis. 2015;**23**:609-618

[13] Agrawal RP, Budania S, Sharma P, Gupta R, Kochar DK, Panwar RB, et al. Zero prevalence of diabetes in camel milk consuming Raica community of north-West Rajasthan, India. Diabetes Research and Clinical Practice. 2007;**76**:290-296

[14] Yadav AK, Kumar R, Priyadarshini L, Singh J. Composition and medicinal properties of camel milk: A review. Asian J. Dairy Food Res. 2015;**34**:83

[15] Lajnaf R. Propriétés techno-fonctionnelles du lait de dromadaire [PhD thesis]. Montpellier, France: University of Montpellier; 2017

[16] Konuspayeva G, Faye B, Loiseau G. The composition of camel

milk: A meta-analysis of the literature data. Journal of Food Composition and Analysis. 2009;**22**:95-101

[17] Alhaj OA, Lajnaf R, Jrad Z, Alshuniaber MA, Jahrami HA, Serag El-Din MF. Comparison of ethanol stability and chemical composition of camel milk from five samples. Animals. 2022;**12**:615

[18] Haddadin MSY, Gammoh SI, Robinson RK. Seasonal variations in the chemical composition of camel milk in Jordan. The Journal of Dairy Research. 2008;**75**:8-12

[19] Lajnaf R, Picart-Palmade L, Cases E, Attia H, Marchesseau S, Ayadi MA. The foaming properties of camel and bovine whey: The impact of pH and heat treatment. Food Chemistry. 2018;**240**:295-303

[20] Lajnaf R, Trigui I, Samet-Bali O, Attia H, Ayadi MA. Comparative study on emulsifying and physico-chemical properties of bovine and camel acid and sweet wheys. Journal of Food Engineering. 2020;**268**:109741

[21] Felfoul I, Lopez C, Gaucheron F, Attia H, Ayadi MA. A laboratory investigation of cow and camel whey proteins deposition under different heat treatments. Food and Bioproducts Processing. 2015;**96**:256-263

[22] Liang LI, Luo Y. Casein and pectin: Structures, interactions, and applications. Trends in Food Science and Technology. 2020;**97**:391-403

[23] Ereifej KI, Alu'datt MH, Alkhalidy HA, Alli I, Rababah T. Comparison and characterisation of fat and protein composition for camel milk from eight Jordanian locations. Food Chemistry. 2011;**127**:282-289

[24] Salami M, Yousefi R, Ehsani MR, Razavi SH, Chobert J-M, Haertlé T, et al. Enzymatic digestion and antioxidant activity of the native and molten globule states of camel α-lactalbumin: Possible significance for use in infant formula. International Dairy Journal. 2009;**19**:518-523

[25] Kappeler S, Farah Z, Puhan Z. Sequence analysis of *Camelus dromedarius* milk caseins. The Journal of Dairy Research. 1998;**65**:209-222

[26] Lajnaf R, Attia H, Ayadi MA. Effect of heat treatments on foaming and physico-chemical properties of bovine and camel sodium caseinate. The Journal of Dairy Research. 2022;**88**:1-5

[27] Lajnaf R, Picart-palmade L, Attia H, Marchesseau S, Ayadi MA. Foaming and air-water interfacial properties of camel milk proteins compared to bovine milk proteins. Food Hydrocolloids. 2022;**126**:107470

[28] Lajnaf R, Zouari A, Trigui I, Attia H, Ayadi MA. Effect of different heating temperatures on foaming properties of camel milk proteins: A comparison with bovine milk proteins. International Dairy Journal. 2020;**104**:104643

[29] Kappeler S. Compositional and structural analysis of camel milk proteins with emphasis on protective proteins. Diss. Techn. Wiss. ETH-Zürich, Nr. 12947, 1998. Ref.: Zdenko Puhan; Korref.: Zakaria Farah; Korref. Gerald Stranzinger; 1998

[30] Hinz K, O'Connor PM, Huppertz T, Ross RP, Kelly AL. Comparison of the principal proteins in bovine, caprine, buffalo, equine and camel milk. The Journal of Dairy Research. 2012;**79**:185-191

[31] Hailu Y, Hansen EB, Seifu E, Eshetu M, Ipsen R, Kappeler S. Functional and technological properties of camel

milk proteins: A review. The Journal of Dairy Research. 2016;**83**:422-429

[32] Omar A, Harbourne N, Oruna-Concha MJ. Quantification of major camel milk proteins by capillary electrophoresis. International Dairy Journal. 2016;**58**:31-35

[33] El-Agamy EI. Bioactive components in camel milk. Bioactive Compounds in Milk and Dairy Products. 2009;**107**:159

[34] Kappeler SR, Farah Z, Puhan Z. 5′-flanking regions of camel Milk genes are highly similar to homologue regions of other species and can be divided into two distinct groups. Journal of Dairy Science. 2003;**86**:498-508

[35] Mohamed H, Johansson M, Lundh Å, Nagy P, Kamal-Eldin A. Short communication: Caseins and α-lactalbumin content of camel milk (*Camelus dromedarius*) determined by capillary electrophoresis. Journal of Dairy Science. 2020;**103**:11094-11099

[36] Felfoul I, Jardin J, Gaucheron F, Attia H, Ayadi MA. Proteomic profiling of camel and cow milk proteins under heat treatment. Food Chemistry. 2017;**216**:161-169

[37] Barzegar A, Yousefi R, Sharifzadeh A, Dalgalarrondo M, Chobert JM, Ganjali MR, et al. Chaperone activities of bovine and camel β-caseins: Importance of their surface hydrophobicity in protection against alcohol dehydrogenase aggregation. International Journal of Biological Macromolecules. 2008;**42**:392-399

[38] Dickinson E. Milk protein interfacial layers and the relationship to emulsion stability and rheology. Colloids Surfaces B Biointerfaces. 2001;**20**:197-210

[39] Romo M, Castellari M, Fartdinov D, Felipe X. Separation of α-lactalbumin enriched fraction from bovine native whey concentrate by combining membrane and high-pressure processing. Food. 2023;**12**:480

[40] Laleye LC, Jobe B, Wasesa AAH. Comparative study on heat stability and functionality of camel and bovine milk whey proteins. Journal of Dairy Science. 2008;**91**:4527-4534

[41] Lajnaf R, Picart-Palmade L, Cases E, Attia H, Marchesseau S, Ayadi MA. The foaming properties of camel and bovine whey: The impact of pH and heat treatment. Food Chemistry. 2018;**240**:295-303

[42] Beg OU, von Bahr-Lindström H, Zaidi ZH, Jörnvall H. The primary structure of alpha-lactalbumin from camel milk. European Journal of Biochemistry. 1985;**147**:233-239

[43] Lajnaf R, Attia H, Ayadi MA. Technological properties and biological activities of camel α-lactalbumin—A review. International Dairy Journal. 2022;**139**:105563

[44] Lajnaf R, Gharsallah H, Jridi M, Attia H, Ayadi MA. Antioxidant and antibacterial activities, interfacial and emulsifying properties of the apo and holo forms of purified camel and bovine α-lactalbumin. International Journal of Biological Macromolecules. 2020;**165**:205-213

[45] Atri MS, Saboury AA, Yousefi R, Chobert J, Haertle T, Moosavi-movahedi AA. Comparative study on heat stability of camel and bovine apo and holo a-lactalbumin. The Journal of Dairy Research. 2010;**77**:43-49. DOI: 10.1017/S0022029909990367

[46] Zennia SSA, Mati A, Saulnier F, Verdier Y, Chiappetta G, Mulliert G, et al. Identification by FT-ICR-MS of Camelus

dromedarius α-lactalbumin variants as the result of nonenzymatic deamidation of Asn-16 and Asn-45. Food Chemistry. 2015;**187**:305-313

[47] Hirayama K, Akashi S, Furuya M, Fukuhara K. Rapid confirmation and revision of the primary structure of bovine serum albumin by ESIMS and Frit-FAB LC/MS. Biochemical and Biophysical Research Communications. 1990;**173**:639-646

[48] El-Hatmi H, Girardet JM, Gaillard JL, Yahyaoui MH, Attia H. Characterisation of whey proteins of camel (*Camelus dromedarius*) milk and colostrum. Small Ruminant Research. 2007;**70**:267-271. DOI: 10.1016/j.smallrumres.2006.04.001

[49] Abd El-Gawad IA, El-Sayed EM, Mahfouz MB, Abd El-Salam AM. Changes of lactoferrin concentration in colostrum and milk from different species. Egyptian Journal of Dairy Science. 1996;**24**:297-308

[50] Kappeler SR, Heuberger C, Farah Z, Puhan Z. Expression of the peptidoglycan recognition protein, PGRP, in the lactating mammary gland. Journal of Dairy Science. 2004;**87**:2660-2668

[51] Ochirkhuyag B, Chobert J, Dalgalarrondo M, Choiset Y, Haertle T. Characterization of whey proteins from Mongolian yak, khainak, and bactrian camel. Journal of Food Biochemistry. 1998;**22**:105-124

[52] Singh H. Heat-induced interactions of proteins in milk. In: Protein & Fat Globule Modifications by Heat Treatment, Homogenization & Other Technological Means for High Quality Dairy Products. Brussels, Belgium: International Dairy Federation; 1995. pp. 86-104

[53] Donato L, Guyomarc'h F. Formation and properties of the whey protein/$\ kappa $-casein complexes in heated skim milk–A review. Dairy Science and Technology. 2009;**89**:3-29

[54] Roefs SPFM, De Kruif KG. A model for the denaturation and aggregation of β-lactoglobulin. European Journal of Biochemistry. 1994;**226**:883-889

[55] Farah Z. Camel Milk Properties and Products. St.Gallen, Switzerland: SKAT. Swiss Centre for Development Cooperation in Technology and Management; 1996

[56] Felfoul I, Lopez C, Gaucheron F, Attia H, Ayadi MA. Fouling behavior of camel and cow milks under different heat treatments. Food and Bioprocess Technology. 2015;**8**:1771-1778

[57] Lajnaf R, Picart-Palmade L, Attia H, Marchesseau S, Ayadi MA. The effect of pH and heat treatments on the foaming properties of purified α-lactalbumin from camel milk. Colloids Surfaces B Biointerfaces. 2017;**156**:55-61

[58] Redington J, Breydo L, Almehdar H, Redwan E, Uversky V. α-Lactalbumin: Of camels and cows. Protein and Peptide Letters. 2016;**23**:1072-1080. DOI: 10.2174/0929866523666160517123738

[59] Lajnaf R, Picart-Palmade L, Attia H, Marchesseau S, Ayadi MA. Foaming and adsorption behavior of bovine and camel proteins mixed layers at the air/water interface. Colloids Surfaces B Biointerfaces. 2016;**151**:287-294

[60] Momen S, Salami M, Alavi F, Emam-Djomeh Z, Moosavi-Movahedi AA. The techno-functional properties of camel whey protein compared to bovine whey protein for fabrication a model high protein emulsion. LWT. 2019;**101**:543-550

[61] Momen S, Salami M, Emam-Djomeh Z, Hosseini E, Sheibani N,

Moosavi-Movahedi AA. Effect of dry heating on physico-chemical, functional properties and digestibility of camel whey protein. International Dairy Journal. 2018;**86**:9-20

[62] Dickinson E. Proteins at interfaces and in emulsions stability, rheology and interactions. Journal of the Chemical Society, Faraday Transactions. 1998;**94**:1657-1669

[63] Borcherding K, Lorenzen PC, Hoffmann W, Schrader K. Effect of foaming temperature and varying time/temperature-conditions of pre-heating on the foaming properties of skimmed milk. International Dairy Journal. 2008;**18**:349-358

[64] Dickinson E. Hydrocolloids at interfaces and the influence on the properties of dispersed systems. Food Hydrocolloids. 2003;**17**:25-39

[65] Zouari A, Lajnaf R, Lopez C, Schuck P, Attia H, Ayadi MA. Physicochemical, techno-functional, and fat melting properties of spray-dried camel and bovine milk powders. Journal of Food Science. 2021;**86**:103-111

[66] Lajnaf R, Gharsallah H, Attia H, Ayadi MA. Comparative study on antioxidant, antimicrobial, emulsifying and physico-chemical properties of purified bovine and camel β-casein. LWT. 2021;**140**:110842

[67] Ellouze M, Lajnaf R, Zouari A, Attia H, Ayadi MA, Vial C. Camel α–lactalbumin at the oil–water interface: Effect of protein concentration and pH change on surface characteristics and emulsifying properties. Colloids Surfaces B Biointerfaces. 2020;**189**:110654

[68] Lam RSH, Nickerson MT. The effect of pH and temperature pre-treatments on the physicochemical and emulsifying properties of whey protein isolate. LWT—Food Science and Technology. 2015;**60**:427-434

[69] Ellouze M, Vial C, Attia H, Ayadi MA. Camel α-lactalbumin at the oil–water interface: Effect of pH and heat treatment on the structure, surface characteristics and emulsifying properties. LWT. 2019;**116**:108550

[70] Ellouze M, Vial C, Attia H, Ayadi MA. Effect of pH and heat treatment on structure, surface characteristics and emulsifying properties of purified camel β-casein. Food Chemistry. 2021;**365**:130421

[71] Zouari A, Marchesseau S, Chevalier-Lucia D, Raffard G, Ayadi MA, Picart-Palmade L. Acid gelation of raw and reconstituted spray-dried dromedary milk: A dynamic approach of gel structuring. International Dairy Journal. 2018;**81**:95-103

[72] Mohamed H, Ayyash M, Kamal-Eldin A. Effect of heat treatments on camel milk proteins–A review. International Dairy Journal. 2022;**133**:105404

[73] Berhe T, Kurtu MY, Ipsen R, Hailu Y, Eshetu M. Coagulation and preparation of soft unripened cheese from camel milk using camel chymosin—A review. East African Journal of Sciences. 2017;**11**:99-106

[74] Qadeer Z, Huma N, Sameen A, Iqbal T. Camel milk cheese: Optimization of processing conditions. Journal of Camelid Science. 2015;**8**:18-25

Chapter 3

Physicochemical Characterization of Mesquite Flour (*Prosopis laevigata*), Particle Size Distribution, Morphology, Isosteric Heat, and Rheology

Sadoth Sandoval Torres, Larissa Giovana Reyes López, Lilia Leticia Méndez Lagunas, Luis Gerardo Barriada Bernal and Juan Rodríguez Ramirez

Abstract

Mesquite pods were dried and milled. The physicochemical properties of mesquite flour were characterized. The pods were dried at 60°C, 15% RH, and 2 m/s airflow. After drying, two types of milling were applied: (1) industrial blade mill and (2) Blender, and the nutritional composition was determined. The sorption isotherms were obtained at 30, 35, 40, and 45°C for a range of water activity of 0.07–0.9. The particle size distribution and the average particle size of the flours were characterized by means of diffraction of blue laser light; furthermore, the morphology was analyzed by (SEM). The powders were also analyzed by DSC. Alveography was applied to study the rheology of the flour. Mesquite powders are highly hygroscopic, and the (GAB) model displays a good description of the experimental data. Flours expose different morphologies depending on the milling technique; a more homogeneous powder was obtained from the industrial blade mill. Rheological characterization indicates that mesquite flour decreases the tenacity and extensibility of the flour mixture. According to DSC, the flours are very stable over a wide temperature range from 0 to 120°C, and the thermograms indicate a transition of proteins affected by high-molecular-weight carbohydrates and moisture content.

Keywords: ethnic food, drying, milling, powder properties, nutritional

1. Introduction

Mesquite (*prosopis spp*) trees grow in dry environmental conditions, counting 44 species in the world [1, 2]. Mesquite tree appertains to genus *Prosopis* and is present in semiarid lands of America [3]. It has been documented that *Prosopis* pods have been food intakes of people from arid and semiarid regions in South America [4], and according to [5], the *Prosopis* pods are highly palatable to humans.

In Mexico, mesquite is found in different regions of the country. In the past, the native people utilized mesquite pods as food, to prepare flour, syrups, and bread [6]; nevertheless, the pod's morphology varied substantially across different regions [7].

According to [1], mesquite pods have a complex morphological structure; the important protein content and the sucrose content reveal a potential for the production of new ethnic foods. As a raw material mesquite pods can be used for baking, snack food, sweetener, gum, and protein concentrate.

Prosopis pallida and *Prosopis juliflora are varieties with especially large and sweet fruits, and* were studied by [8]. The authors explored different applications for every component of the fruit (exocarp, mesocarp (pulp), and endocarp), the episperm, endosperm, and cotyledon of the seed. They affirm sucrose is the main sugar component in the pulp; galactomannan is the most important polysaccharide in the endosperm, and glutamic acid, arginine, aspartic acid, leucine, proline, and serine were identified in the seed cotyledon. In the pulp, vitamin C, nicotinic acid, and calcium pantothenate were identified. In other study [9], authors characterize the phenolic antioxidants occurring in the pod mesocarp flour of Chilean *Prosopis*; they conduct an HPLC-MS/MS analysis identifying the presence of eight anthocyanins and 13 phenolic compounds including flavonol glycosides, C-glycosyl flavones, and ellagic acid derivatives. The antioxidant activity and the phenolic composition of this product reveal its potential as a functional food.

The structural and functional properties of *P. alba, P. chilensis, and P. flexuosa* were assessed by other author [10]. The flours were characterized by granulometric analyses, water absorption, oil absorption, solubility, and color. Drying and milling process allows ultrastructural changes, modifying the membranes of proteins and changing their capacity to absorb water. The values of solubility reveal flours can be used for the elaboration of liquid foods and candies. According to [2], *Prosopis* mesocarp flour contributes to the browning, color, aroma, and flavor of baked products.

In other work [11], the authors studied the fruits of *Prosopis alba* and *P. pallida*. A drying process was applied at 60°C for 60 h, then a hammer milling process was used. The findings reveal proteins, calcium, iron, dietary fiber, and sugars as the principal constituents of the pulp. The efficiency of the milling and sieving for *P. alba* and *P. pallida* were 54.5% and 55%, respectively. The total sugar content was higher in *P. alba* than *P. pallida* and protein content was higher in *P. pallida* than *P. alba*.

In another study, the authors investigated the particle size, morphology, rheology, physicochemical, and mineral composition of *Proposis julifrora* [3]. Drying at 60°C was applied, after that the pods were analyzed. Sieves with a meshing of 32–150 were used. The flour exhibited an important concentration of fibers, calcium, and phosphorous. The results of rheology indicated that mesquite flour is suitable for cookie production.

In other work, the genotoxicity of *prosopis* flour was addressed [4]. The authors affirm sucrose constitutes the main sugar in flours obtained from *P. alba* and *Prosopis nigra*. *Prosopis* extracts did not reveal any mutagenic effect with and without metabolic activation. The authors conclude *Prosopis* flour is a rich source of antioxidant compounds that could avoid pathologies related to oxidative stress.

For the elaboration of bakery products and/or confectionery, it is indispensable the use of flours with the ideal characteristics that can satisfy the culinary necessities and that in turn can be conserved [10]. Currently, the use of mesquite pods in the food industry is uncommon [2, 5]. Then, the aim of this work was to assess the physicochemical and rheological characteristics of mesquite flours (*Prosopis laevigata*) harvested in Oaxaca State (Southern of Mexico) in order to develop foods with important nutritional value without gluten and take advantage of the agro-food resources of semiarid zones. The particle size distribution, the particle morphology, the isosteric heat of sorption, and the thermal stability were also studied.

2. Materials and methods

Pods of *P. laevigata* were harvested between April and August 2016 in the community of Santiago Sulchiquitongo (Oaxaca, Mexico). Three stages of maturation were identified [12]. Pods in stage three of maturity were used for drying. The drying process of pods was performed using a convective tunnel dryer [13]. The drying conditions were as follows: an airflow at 60°C, with a relative humidity of 15%, and an air velocity of 2.0 m/s. After drying, pods were stored in a desiccator.

The pods reached a final moisture content of 0.12 g of water/g dry matter. Once dried, the mesquite was milled by implementing two techniques; 1) an Osterizer blender, model 465–15 for 20 seconds and 2) a mill blade pulverizer Model HC-2000Y, for 20 seconds. The mesquite powder was passed through # 60 (0.250 mm) and # 80 (0.177 mm) sieves and each mill was stored in low-density polyethylene bags in a vacuum desiccator for 24 hours. A more homogeneous material was obtained from the mill blade pulverizer, then, powders from this milling technique were subsequently analyzed.

2.1 Chemical characterization of flours

The chemical-proximal and nutritional composition of the flours were obtained. The moisture content, the total raw protein content, the reducing and direct sugars, the total fat extraction, the raw fiber, and the ash were determined by methods published in [12].

2.2 Particle size distribution

The size distribution of a particulate product is dependent on the shapes of its particles [13, 14]. The particle size distribution determines the critical chemical and physical properties of particulate systems [15]. Particle size induces many properties of powder materials and is a significant indicator of quality and performance. For this reason, the particle size distribution of mesquite flours was analyzed using the principle of blue laser light diffraction measurement [16]. For this purpose, we use a Microtrac Blueray M3551-1 W-BU00 in a humid medium, with a measuring range of 10 nm up to 2000 microns. The method is presented in [12].

2.3 Morphology of flour particles

Images from samples of mesquite flour were obtained. A scanning electron microscope (SEM) JEOL brand, model JIB-4601F, with a spatial resolution of 1.2 nm, a focused ion beam, and a digital camera (CIIDIR-Oaxaca, Mexico), was used in this work. The flour samples were placed in small graphite plates and introduced into the SEM vacuum chamber. In our analysis, a secondary electron detector E-T (Everhart-Thornley) and a backscattered electron detector were used. A range of magnification from 50x to 2500x was used for the images.

2.4 Sorption isotherms

According to [17], powder processing must be conducted under controlled relative humidity and temperature in order to enhance the storage, handling, and processing. As the relative humidity of the surrounding air is increased, powders tend to absorb water, which may form liquid bridges between powder particles and result in greater powder cohesion. The sorption isotherms of mesquite powders (*Prosopis Laevigata*) were assessed by the gravimetric static method with water activities ranging from 0.07 to 0.97 at four temperatures: 30, 35, 40, and 45°C. The salts used in this work were the following: NaCl, $MgCl_2 \cdot 6H_2O$, KOH, KCl, KI, K_2SO_6, and $Mg(NO_3)_2 \cdot 6H_2O$ [18, 19] The details of this method are presented in [12]. Experimental data was fitted to the (GAB) (Guggenheim-Anderson-Deboer) model (Eq. (1)). The theoretical fundamental for the GAB sorption isotherm is the assumption of localized physical adsorption in multilayers without lateral interactions [20]. The parameters for this model were estimated by implementing the equation in excel (GRG nonlinear).

$$\mathrm{Xeq} = \frac{\mathrm{Xm} \cdot \mathrm{C} \cdot \mathrm{k} \cdot \mathrm{aw}}{(1 - \mathrm{k} \cdot \mathrm{aw}) \cdot (1 - \mathrm{k} \cdot \mathrm{aw} + \mathrm{C} \cdot \mathrm{k} \cdot \mathrm{aw})} \tag{1}$$

where *Xeq* is the equilibrium moisture content (g H_2O/g dry basis), X_m is the monolayer moisture content (g H_2O/g dry basis), and C is a heat-related constant of the monolayer, *k* is a constant related to the sorption heat of the multilayer, and *aw* is the water activity.

2.5 Isosteric heat of sorption

The estimation of energy consumption during drying needs the knowledge of the enthalpy of water sorption in the entire range of moisture contents. Certainly, the use of the enthalpy of vaporization of pure water can give inaccurate results [21]. The isosteric heat of sorption is a useful expression, notably in the design of drying operations, as heats of sorption rise well in excess of the heat of vaporization of water as food is dried to low moisture contents [22]. The net isosteric heat of sorption measures the binding energy of the forces between the water vapor molecules and the solid phase [23]. It gives information for the comprehension of the sorption mechanism [20]. The gap between the amount of energy necessary to remove water from the flour and the amount of energy needed for normal water vaporization is defined as the net isosteric heat of sorption. The isosteric heat of sorption (*Qst*) was estimated by using the equation derived from Clausius Clapeyron [22–24]; in order to calculate the enthalpy change associated with the sorption process (Eq. (2)).

$$\left[\frac{\partial \ln (a_w)}{\partial(1/T)}\right]_{CHE} = -\frac{Q_{st} - \lambda}{R} = -\frac{q_{st}}{R} \tag{2}$$

where q_{st} is the isosteric heat of sorption, R is the universal constant of the gases (0.00831 Joules/K·mol), a_w is the water activity, T is the temperature (K), and λ is the latent heat of vaporization of pure water at room temperature. For isosteric heat of sorption, we used the experimental information of the sorption isotherms.

2.6 Differential scanning calorimetry (DSC)

The scanning calorimetry provides a direct estimate of the overall enthalpy change of transitions without requiring knowledge of the thermodynamic mechanism; moreover, the sample preparation is minimal [25]. Four samples of mesquite flours (*P. laevigata*) previously conditioned at a relative humidity of 7%, 32%, 51%, and 67% were prepared for a DSC analysis (TA Instruments, model Q2000). 17 mg of flour was placed inside hermetic aluminum capsules and sealed by a press. The experimental conditions consisted in running the samples at an initial temperature of 0.0°C, followed by a heating rate of 2°C/min up to an end temperature of 250°C. The thermograms were analyzed by using the TA Instruments DSC software.

2.7 Alveography

Alveograph method is useful to estimate the potential performance of flours [26]. The Alveograph test provides a test sample of dough, which, under air pressure, forms a bubble. The test recreates the deformation of a dough when is subjected to carbondioxide during fermentation [27].

Alveographic characteristics were analyzed for three mixtures of wheat/mesquite flour. Experiments were conducted in triplicate in alveographic equipment (Chopin Technologies, France), following the AACC Method 54–30.02 [28], quantifying the following parameters: tenacity (P), extensibility (L), and dough deformation energy (W).

The tenacity is the capacity to resist deformation, extensibility is the maximum volume of air that the bubble is able to enclose, and deformation energy corresponds to the dough baking strength.

3. Results and discussion

The pods show different shades of color during maturity; these changes are accompanied by variations in the organoleptic and physical properties. According to [12], pods reveal three stages of maturity: pods in stage 1 showed a green coloration, in stage 2 the pods were brighter with a reddish coloration, and in stage 3 the pods increased in brightness compared to the previous stages, reddish and yellow coloration was still observed, cream color is characteristic of this last stage. For the study, we used exclusively pods in stage 3 of maturation.

3.1 Chemical composition of flours

The nutritional composition of mesquite flours is shown in (**Table 1**). The nutritional compositions of mesquite flours present a complete composition of macronutrients. Some differences according to the type of milling were observed, the fiber content was higher for milling with a blade mill; however, the proteins decreased slightly. Both flours are suitable for use as complement flour or as a natural supplement.

Components	Blender (1)		Blade (2)	
	Average (g/100 g)	Standard deviation	Average (g/100 g)	Standard deviation
Energy content (kcal/100 g)	170.97	0.07	198.72	0.50
Carbohydrates (g)	24.27	0.09	26.18	0.08
Sugars (g)	7.48	0.03	10.18	0.02
Proteins (g)	12.4	0.08	11.77	0.12
Fats (Lipids) (g)	2.16	0.03	2.90	0.06
Fiber (g)	16.9	0.2	17.25	0.11
Ashes (g)	3.12	0.01	3.45	0.05
Humidity (g water/g dry matter)	0.10	0.01	0.10	0.01

Table 1.
*Nutritional composition of mesquite flour (*Prosopis laevigata*) obtained by two different grindings.*

3.2 The particle size distribution

In **Figure 1** we show the size distribution of the flours. During the procedure, the aggregates were dispersed with the help of the sample dispersion accessory, and a power of 30 watts was applied for 30 s.

For the flour obtained from the blender (**Figure 1a**), two oblations were identified. Particles of 1.291 microns with 95% up to 657 microns and a cumulative 10% of particles of 28.93 microns were measured.

For the blade mill (**Figure 1b**), the particle size distribution showed a smooth and unique Gaussian distribution. The sample was dispersed very nicely with the simple agitation of the circulatory without forming aggregates. The fine particles are presented from 28.53 microns with a cumulative 10% of particles of 64.23 microns. There are particles as large as 497 microns, with a cumulative 95% of 302.5 microns. The powders showed a homogeneous distribution with adequate particle size. Mesquite flour can be used for the elaboration of baking and confectionery products.

3.3 Morphology of flour particles

The (SEM) images of flour particles are shown in **Figure 2**. The two methods of milling produce different morphologies of powders. Diverse shapes and sizes of mesquite particles with irregular surfaces are identified, and the smooth and striated parts are shown.

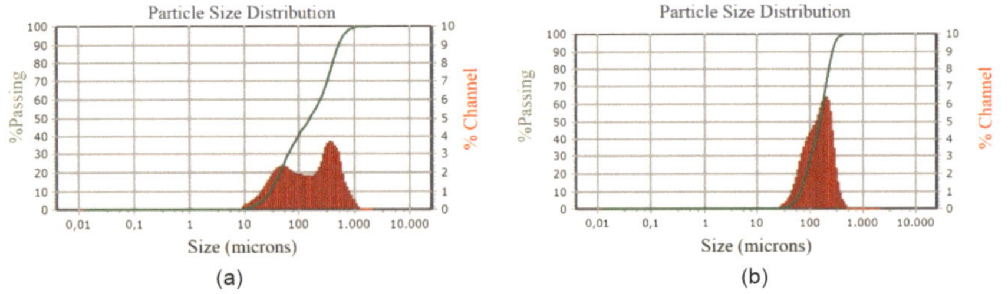

Figure 1.
(a) Particle size distribution – Blender; and (b) Particle size distribution – Blade mill pulverizer.

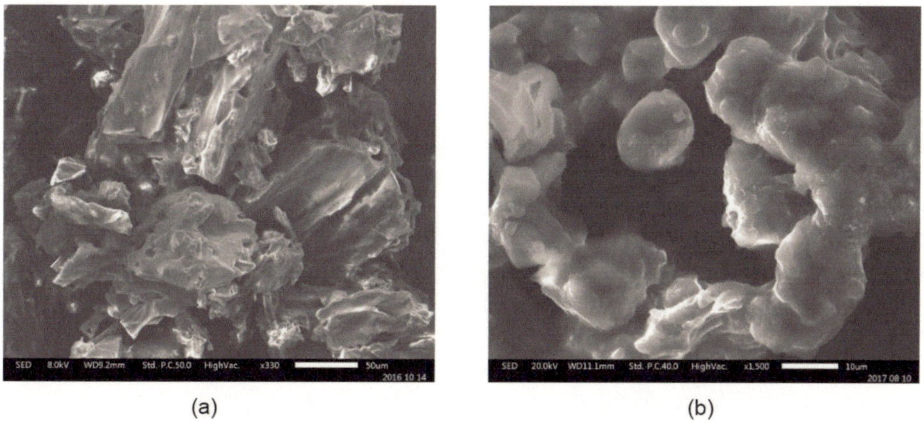

Figure 2.
(a) Micrographs of mesquite flour (Blender), 330×; and (b) Micrograph of mesquite flour (Blade mill), 1500 ×.

Figure 2a shows the morphology of powders obtained from the blender. It shows a slight agglomeration and striated parts forming a tortuous, irregular, and rocky agglomerate. It can be observed the presence of some cavities that can allow the moisture adsorption. **Figure 2b** shows the morphology of powders obtained by blade mill. It reveals the surface of a particle with a rounded structure, better defined, with smooth surfaces without forced cuts.

3.4 Sorption isotherms

The experimental information and the simulations of the GAB model at 30, 35, 40, and 45°C are shown in **Figure 3**, which display a Type II isotherm shape, and indicate a likely small adsorption force in the monolayer [29, 30]. This type of isotherm has been presented in materials containing fibers and polysaccharides of cereals (wheat,

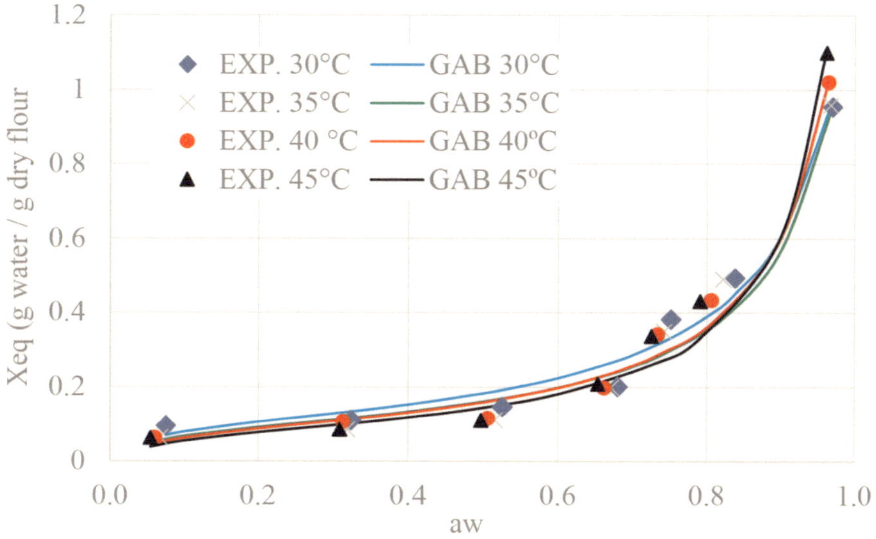

Figure 3.
Sorption isotherm of mesquite flour (Prosopis laevigata) and simulated GAB model.

Model	Parameters	30 °C	35°C	40°C	45°C
GAB	Xwa (g water/g dry matter)	0.1039	0.0905	0.0894	0.0824
	C	22.1997	19.9075	16.6142	15.2432
	k	0.9219	0.9355	0.9475	0.9656
	r^2	0.9898	0.9772	0.9817	0.9773
	s	0.0555	0.0741	0.0592	0.0703

Xwa = Monolayer value; C and k = constants for the model; r^2 = correlation coefficient; and s = standard error.

Table 2.
Estimated parameters for the GAB model.

rice), tubers (potato, cassava), proteins (soybean, maize), and starches, these components are present in the mesquite flour. In the experimental curves, it was observed that the equilibrium moisture content notably increases for aw = 0.67–0.97. The GAB model correctly represented the experimental data for all experimental conditions, giving a %Error of 10–15%. **Table 2** shows the parameters estimated from the GAB model.

According to [20], parameter C indicates the strength of binding of water to the primary binding sites. The larger the C, the stronger water is bound in the monolayer, and the larger the difference in enthalpy between the monolayer molecules and multilayer molecules. In the case of parameter K, when it approaches one (our case), there is almost no divergence between multilayer molecules and liquid molecules. In that case, the water molecules beyond the monolayer are not structured in a multilayer, but have the same characteristics as the molecules in the bulk liquid, as discussed by [31].

3.5 Isosteric heat of sorption

Figure 4 shows the evolution of isosteric heat (Qst) versus the moisture content (Xw) of mesquite flour (*Prosopis laevigata*). We observe Qst decreased rapidly with the increase of Xw; being the lowest value of Qst at 47.69 kJ/mol at 0.25 of Xw. This situation indicates a low availability of active sites and liaison forces on the surface of the powder [23]. When the moisture content is 0.15 g of water/g of dry matter, the heat needed to evaporate the water from mesquite flour would be 57.03 kJ/mol., without affecting the stability of the powder. The net isosteric heat of sorption estimates the binding energy of the forces between the water vapor molecules and the solid phase. It allows a better comprehension of the sorption mechanism.

Figure 4 depicts positive quantities, manifesting the endothermic behavior of desorption. The isosteric heat of sorption decreases as moisture content increases. This fact refers to the intermolecular attraction forces between sorptive sites and water vapor.

The higher the moisture content, the less energy is necessary to remove water molecules from the flour. As drying continues sorption will perform at active sites demanding higher interactive energies [20].

3.5.1 Differential scanning calorimetry (DSC)

Since food powders are complicated mixtures of compounds, it is regularly difficult to identify their phase's transitions accurately. **Figure 5** shows the DSC curves for

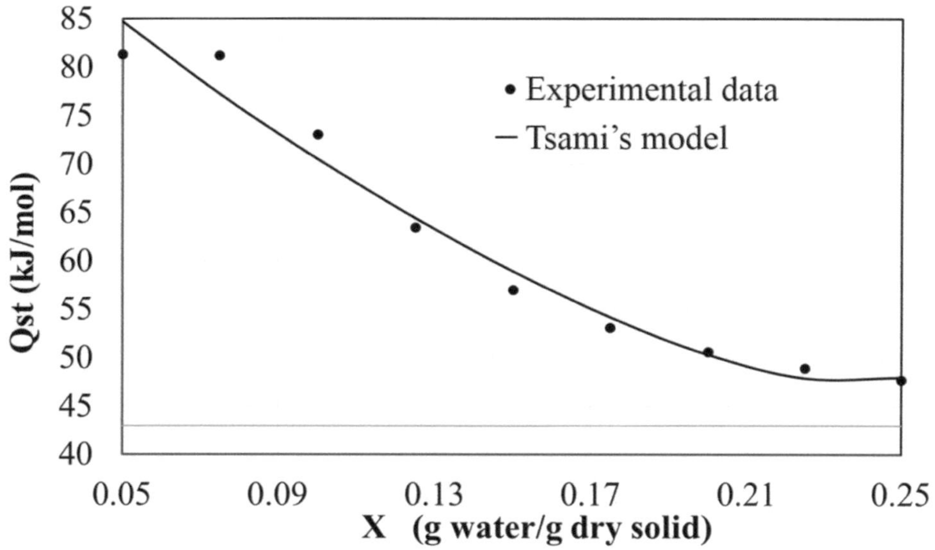

Figure 4.
Isosteric heat of sorption.

mesquite flours exposed at four RH (6.3, 31.8, 48.5, and 66.1%). If it is true that flour contains protein, the heat denaturation temperatures of proteins *in solution* are normally below 100°C; nevertheless, proteins become stable toward heat when the moisture content is low. A clear example of moisture effect on protein denaturation is published in [32]. A DSC thermogram for wheat flour is presented by [17]; in this work, a crystallization peak was observed near 190°C, which is referred by the authors as a decomposition of the flour. In our work, the glass transition (Tg) of the mesquite flour was not observed, due to possible flexibility and mobility of the glucose and fructose chains, provoked by the increase of moisture content of the powders when exposed at different RH (6.3, 31.8, 48.5, and 66.1%). According to [33], the glass transitions of high molecular weight carbohydrates and proteins arise well above 100°C and approach thermal decomposition temperatures of the food powder. Moreover, the plasticization effect of water leads to depression of the glass transition temperature causing noticeable changes in the physicochemical and crystallization properties of the material.

The transitions shown in **Figure 5** display a first-order behavior. The transition phase that was observed in all the flour samples was the crystallization peaks (Tc) from 140–157°C. Likewise, the heat flow of the endothermic transitions increased as the moisture of the flours increased, corroborating the effect of the relative humidity in the modification of the structure of the flours. It is also well-known sugars affect the protein thermal properties [25]. The thermograms of mesquite flour showed significant endotherms in the range of 130–180°C, being associated with a melting of the simple sugars present in the flour.

According to Barba de la Rosa et al. [34], sugars such as glucose, maltose, L-arabinose, and sucrose are found in high proportion in mesquite pods, so the transitions for this food can be related to a binary water-carbohydrate system.

The thermograms in **Figure 5** show the transition of proteins affected by high molecular weight carbohydrates and moisture content of powders.

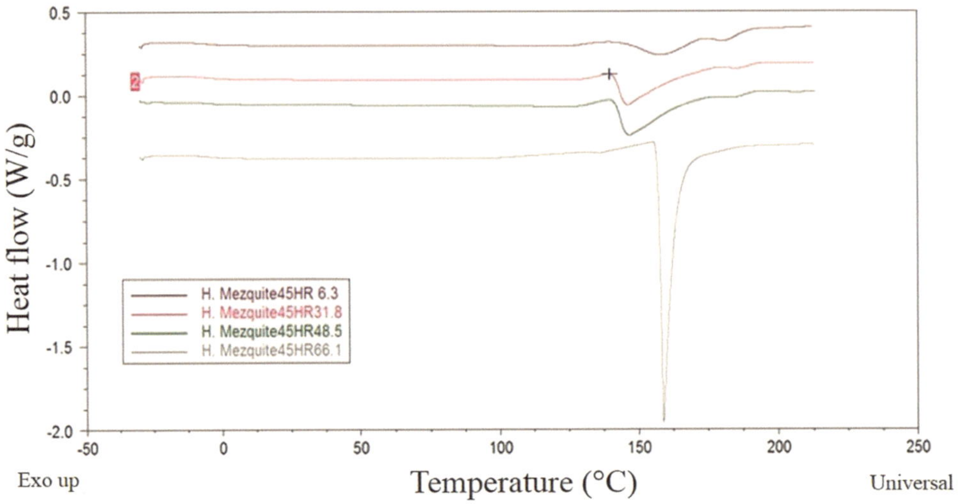

Figure 5.
*DSC curves of mesquite flours (*Prosopis laevigata*).*

3.6 Alveography

The water absorption of mesquite flour can significantly affect the rheological properties of a mixture. According to our results (**Table 3**), the dough's tenacity ranged 18–83 mmH_2O, decreasing with the increase of mesquite flour content of 0–15% (wt). According to [35], tenacity values for standard wheat quality range 60–80 mm H_2O, and very good wheat quality 80–100 mm H_2O, then mesquite flour decrease the quality of the mixture, for this reason, it should be mixed at low concentrations for baking applications.

Extensibility characterizes the average length of the alveogram from the point at which the bubble starts to inflate to the point at which the bubble breaks. Extensibility (L) decreases as mesquite content increases, so mesquite flour impacts the handling properties of the dough. According to [3], when the concentration of mesquite flour is increased, tenacity and extensibility are reduced, and this fact can be explained by the weakening of glutenin protein (a protein responsible for elasticity and extensibility), and also due to a lower water absorption due to the high fiber content of mesquite flour. The higher the addition of mesquite flour with wheat flour, the dough may have a lower capacity to retain the gas generated during fermentation and rising of the bread. The energy of deformation (W) ranged 2.6–4.2, indicating that as mesquite

Mixture % wt Wheat flour + % wt Mesquite flour	Tenacity (P) Average ± SD	Extensibility (L), Average ± SD	Deformation energy (W), Average ± SD
100% + 0%	83.0 ± 6.0415	74.6 ± 7.8930	4.20 ± 3.2985
95% + 5%	35.4 ± 2.6076	81.0 ± 6.2449	2.63 ± 2.5467
90% + 10%	22.4 ± 5.0793	50.4 ± 32.9135	2.74 ± 0.8354
85% + 15%	18.4 ± 3.1304	19.0 ± 5.3385	2.67 ± 2.0255

Table 3.
Results of the alveography.

proportion increases the baking strength decreases. W is frequently referred to as flour strength, dough strength, baking strength, or flour protein strength [36]. W is one of the industrially most applied alveograph parameters, as it is used for prediction of processing behavior of flour cultivars [35]. For example, according to [37] bread flours are characterized by larger W values compared to biscuit flours. W is positively related to the water absorption of the flour [38]. According to our results, an increase in mesquite flour content reduces water absorption, then, this effect should be considered on the type of application, either in baking or as biscuits.

4. Conclusions

Dried pods were milled by two milling techniques. The two techniques of milling produce different morphologies of powders. Particles from the blade mill were more homogeneous, they showed a smooth and unique Gaussian distribution, and SEM images reveal a rounded structure, better defined, without rocky parts nor forced cuts. The flour from blade mill has an adequate particle size to be used as flour for baking and confectionery products. Flours from the two different grindings have an important content of protein, sugar, and fibers. Mesquite flours showed a type II isotherm for the three experimental temperatures, indicating a possible small adsorption force in the monolayer. Flours are notably hygroscopic, and this phenomenon is related to the sugar content and the powder's microstructure. The isosteric heat reveals the endothermic behavior of desorption and the energy required to remove the water molecules from the powder.

The calorimetric data of the flours showed thermal stability in a wide temperature range; however, for temperature > 130°C crystallization peaks (Tc) were observed (from 140–157°C), which show the transition of proteins affected by sugars and moisture content.

Our results confirm that, as the mesquite flour proportion increases in the mixture, the flour became poorer. Tenacity values for standard wheat quality range 60–80 mm H_2O, whilst our mixtures showed lower values. According to the values for extensibility, the presence of mesquite flour affects the handling properties of the dough. The values of deformation energy show mesquite flour develop a weak baking strength, then it should mix at low concentrations for baking applications since an increase in mesquite flour content reduces water absorption. The nutritional composition of the flour releases useful attributes for many applications in food industry, however, this information should be carefully studied, depending on the application of mesquite flour, either in baking or as biscuits.

Acknowledgements

Authors express special thanks to Conacyt for the scholarship granted to Larissa G. Reyes López and Daniel López Cravioto, and the Instituto Politénico Nacional (Mexico) for SIP funding 20161016, 20170755, 20180678, and 20195013.

The authors are particularly grateful for the assistance given by Administración Profesional de Servicios Xoluciona S.A. de C.V. for the technical assistance in the analysis of particle size distribution and the use of Microtrac Blueray M3551-1 W-BU00.

Funding

This work was supported by the Instituto Politécnico Nacional and CONACYT.

Author details

Sadoth Sandoval Torres[1*], Larissa Giovana Reyes López[1],
Lilia Leticia Méndez Lagunas[1], Luis Gerardo Barriada Bernal[2] and
Juan Rodríguez Ramirez[1]

1 Instituto Politécnico Nacional, CIIDIR Unidad Oaxaca, Oaxaca, México

2 Consejo Nacional de ciencia y Tecnología. Instituto Politécnico Nacional, Oaxaca, México

*Address all correspondence to: ssandovalt@ipn.mx

References

[1] Meyer, D. 1984. Processing, Utilization and Economics of Mesquite Pods as a Raw Material for the Food Industry. Zurich: PhD. Thesis. ETH Zurich

[2] Felker P, Takeoka G, Dao L. Pod mesocarp flour of north and south american species of leguminous tree prosopis (Mesquite): Composition and food applications. Food Reviews International. 2013;**29**:49-66. DOI: 10.1080/87559129.2012.692139

[3] Gusmao RP, Cavalcanti Mata ME, Martins Duarte ME, Souza Gusmao TA. Particle size, morphological, rheological, physicochemical characterization and designation of minerals in mesquite flour (Proposis julifrora). Journal of Cereal Science. 2016;**69**:119-124. DOI: 10.1016/j.jcs.2016.02.017

[4] Cardozo M, Ordoñez RM, Zampini IC, Cuello AS, Dibenedetto G, Isla MI. Evaluation of antioxidant capacity, genotoxicity and polyphenol content of non conventional foods: Prosopis flour. Food Research International. 2010;**43**(5):1505-1510. DOI: 10.1016/j.foodres.2010.04.004

[5] Felker P. Unusual physiological properties of the arid addapted tree legume Prosopis and their applications in developing countries. In: Smith ED, editor. Perspectives in Biophysical Plant Ecophysiology: A Tribute to Park S. Mexico D.F., México: Nobel Universidad Nacional Autónoma de México; 2009. pp. 221-255

[6] Figueiredo AA. Mesquite: History, composition and food uses. Food Technology. 1990:**40**:118-128

[7] Felker P, Grados N, Cruz G, Prokopiuk D. Economic assessment of production of flour from Prosopis alba and P. pallida pods for human food applications. Journal of Arid Environments. 2003;**53**:517-528. DOI: 10.1006/jare.2002.1064

[8] Grados N, Cruz G. New Approaches to Industrialization of Algarrobo (Prosopis pallida) Pods in Peru. Prosopis: Semiarid Fuelwood and Forage Tree; 1996. pp. 25-42

[9] Schmeda-Hirschmann G, Quispe C, Soriano Maria Del Pilar C, Theoduloz C, Jiménez-Aspée F, Jorgelina Pérez M, et al. Chilean prosopis mesocarp flour: Phenolic profiling and antioxidant activity. Molecules. 2015, 2015;**20**(4): 7017-7033. DOI: 10.3390/molecules 20047017

[10] Mom M. Caracterización estructural y propiedades funcionales de las harinas de los frutos de Prosopis alba Griseb., P. chilensis (Molina) y P. flexuosa DC. Desarrollo de un proceso de secado, molienda y mezcla para optimizar la calidad del producto. Thesis. Buenos Aire, Argentina: PhD. Universidad de Buenos Aires; 2012

[11] Prokopiuk D, Grados N, Garro O, Chiralt A. Estudio comparativo entre frutos de Prosopis Alba y Prosopis pallida. Multequina. 2000:**9**:35-45

[12] Sandoval TS, Reyes López LG, Méndez Lagunas L, Rodríguez Ramírez J. Physicochemical Charcaterization of Mezquite Flours. Valencia, Spain: International Drying Symposium; 2018

[13] Sandoval Torres S, Cravioto DL, Ramírez JR, Méndez Lagunas LL, Bernal LGB, Aquino González LV. Drying technology of mesquite pods (Prosopis laevigata) and

microestructural insights. British Food Journal. 2020;**122**(9):2953-2963. DOI: 10.1108/BFJ-07-2019-0487

[14] Califice A, Michel F, Dislaire G, Pirard E. Influence of particle shape on size distribution measurements by 3D and 2D images analyses and lasser diffraction. Powder Technology. 2013;**237**:67-75. DOI: 10.1016/j.powtec.2013.01.003

[15] Basim G, Khalili M. Particle size analysis on wide size distribution powders; effects of sampling and characterization technique. Advanced Powder Technology. 2015, 2015;**26**: 200-207. DOI: 10.1016/j.apt.2014.09.009

[16] ISO13320. Particle Size Analysis. International Organization for Standardization: Laser diffraction methods; 2009

[17] Teunou E, Fitzpatrick JJ. Effect of relative humidity and temperature on food powder flowability. Journal of Food Engineering. 1999;**42**(2):109-116. DOI: 10.1016/S0260-8774(99)00087-4

[18] Greenspan L. Humidity fixed points of binary saturated aqueous solutions. Journal of Research of the National Bureau of Standards Section A: Physics and Chemistry. 1977;**81**(1):89-96

[19] Sahin S, Sumnu SG. Physical Properties of Foods. 1st ed. USA: Springer; 2006

[20] Quirijns E, van Boxtel AJ, van Loon WK, van Straten G. Sorption isotherms, GAB parameters and isosteric heat of sorption. Journal of the Science of Food and Agriculture. 2005;**85**(11): 1805-1814. DOI: 10.1002/jsfa.2140

[21] Iglesias H, Chirife J. Handbook of Food Isotherms: Water Sorption Parameters for Food and Food Components. 1st. ed. New York: Academic Press; 1982

[22] McLaughlin C, Magee TRA, T. R. The determination of sorption isotherm and the isosteric heats of sorption for potatoes. Journal of Food Engineering. 1998;**35**(3):267-280. DOI: 10.1016/S0260-8774(98)00025-9

[23] Tsami E. Net isosteric heat of sorption in dired fruits. Journal of Food Engineering. 1991;**14**(4):327-335. DOI: 10.1016/0260-8774(91)90022-K

[24] Rao M, Rizvi SS, Datta AK, Ahmed J. Engineering Properties of Foods. 4th ed. Boca Raton: CRC Press; 2014

[25] Farkas J, Mohácsi Farkas C. Application of differential scanning calorimetry in food research and food quality assurance. Review. Journal of Thermal Analysis. 1996:**47**:1787-1803

[26] Li J, Gary GH, Zhengxing C, Chung A-L, Gehring K. Studying the effects of whole-wheat flour on the rheological properties and the quality attributes of whole-wheat saltine cracker using SRC, alveograph, rheometer, and NMR technique. LWT - Food Science and Technology. 2014;**55**(11):43-50. DOI: 10.1016/j.lwt.2013.07.022

[27] Chopin T. Alveolab. n.d. Available from: https://chopin.fr/en/home.html

[28] AACC. Physical Dough Tests. Method 54-30.02. Approved methods of analysis. 11th ed2000

[29] Anderson R. Modification of the BET equation. Journal of the American Chemical Society. 1946;**1946**(68): 686-691

[30] Charreau A, Cavaillé RS. Théorie et calculs. Techniques de l'Ingénieur. 1995: 1-23

[31] Pérez Alonso C. Thermodynamic analysis of the sorption isotherms of pure and blended carbohydrate polymers. Journal of Food Engineering. 2006;**77**(4):753-760. DOI: 10.1016/j.jfoodeng.2005.08.002

[32] Kitabatake N, Tahara M, Doi E. Thermal denaturation of soybean protein at low water contents. Agricultural and Biological Chemistry. 1990;**54**:2205-2212. DOI: 10.1080/00021369.1990.10870318

[33] Roos Y. Phase transitions and structure of solid food matrices. Current Opinion in Colloid & Interface Science. 1998;**3**(6):651-656. DOI: 10.1016/S1359-0294(98)80095-2

[34] Barba De La Rosa AP, Frías Hernández JT, Olalde Portugal V, González Castañeda J. Processing, nutritional evaluation, and utilization of whole mesquite flour (Prosopis laevigata). Journal of Food Science. 2006;**71**(4):315-320. DOI: 10.1111/j.1750-3841.2006.00001.x

[35] Bordes J, Branlard G, Oury FX, Charmet G, Balfourier F. Agronomic characteristics, grain quality and flour rheology of 372 bread wheats in a worldwide core collection. Journal of Cereal Science. 2008;**48**(3):569-579. DOI: 10.1016/j.jcs.2008.05.005

[36] Jødal AS, Larsen KL. Investigation of the relationships between the alveograph parameters. Scientific Reports. 2021;**11**(1):5349. DOI: 10.1038/s41598-021-84959-3

[37] García-Álvarez J, Salazar J, Rosell CM. Ultrasonic study of wheat flour properties. Ultrasonics. 2011;**51**:223-228. DOI: 10.1016/j.ultras.2010.08.006

[38] Dexter JE, Preston KR, Martin DG, Gander EJ. The effects of protein content and starch damage on the physical dough properties and bread-making quality of Canadian durum wheat. Journal of Cereal Science. 1994;**20**:139-151. DOI: 10.1006/jcrs.1994.1054

Chapter 4

Development and Optimization of Flakes from Some Selected Locally Available Food Materials

Samuel Tunde Olorunsogo and Bolanle Adenike Adejumo

Abstract

Flakes are one of the most popular ready-to-eat breakfast cereals meals. Most traditional instant breakfast meals are from mono-cereals. This work aims to develop, characterize and optimize value-added instant cereal breakfast flakes using flours of rice, sorghum, and soybean. A three-component constrained optimal (custom) mixture experimental design was employed for the formulation. The formulation design constraints were: rice flour (30%–35%), sorghum flour (20%–25%), and soybean flour (5%–10%). Other ingredients were water (19%), sugar (8%), malt (2%), egg (3%), sweet potato (3%), ginger (2%) and moringa seed powder (3%). The formulated samples were analysed and evaluated based on standard procedures for quality characteristics. Numerical optimization gave the optimal product's overall desirability index of 0.519 obtained from 31.9 % rice flour, 22% sorghum flour, and 6.05% soybean flour; with quality properties as follows: 3.67% moisture content, 3.18% fat content, 3.08% ash content, 1.44% crude fibre, 30.0% crude protein, 58.6% nitrogen-free extract, 384 kcal energy value, and 7.28 overall acceptability. The result of the study showed that the nutritional qualities of cereal flakes can be improved through food-to-food composite formulations, employing numerical optimization technique.

Keywords: instant flakes, multigrain, breakfast cereals, optimization, development

1. Introduction

The food manufacturing process is the transformation of raw ingredients into edible end products for human consumption. It has been a popular method of producing convenient and accessible food since ancient times. Knowing the formulation and processing profile of a food or beverage and how it quantitatively relates to consumer perceptions opens up a world of development, quality, and marketing opportunities for a food manufacturer. Methodical exploration of product features, known commonly as "response surface methodology" (RSM) is vital to manufacturing of quality products.

Cereals are the basis of many staple foods and have been used in flaking for over a century. They provide over half of the dietary energy globally and are a major source of carbohydrates in the diet. Most traditional instant breakfast meals are produced from mono-cereals and these mono-cereals consist of carbohydrate as its major constituent. The nutritional and energy level gotten from the consumption of these mono-cereals is minimal. Besides, the production involves many processes and during these processes, nutrient losses occur therefore reducing its nutritional content. Major constituents of breakfast cereals are whole or broken cereal kernels (flaked, cracked) or ground (flours or meals). Importantly, new technology, like extrusion, enabled higher production rates and lowered manufacturing costs, but ended up having an impact on where cereals are today [1, 2].

Flakes are convenient and relatively shelf stable breakfast cereals, primarily produced from corn, wheat, rice, and/or oats, and processed with added flavor and fortified with vitamins and minerals. Flaking process is a relatively simple process of cooking fragments of cereal grains (or in some cases whole grains) with water, flattening the particles between large steel rollers and toasting the resultant flake at high temperature. In flaking the starch in gelatinized and probably slightly hydrolyzed. The particle then undergoes dextrinization and caramelization.

There are other grains that can be used but are presently unexploited. Soybean contains about 40% protein; it is higher than other legumes in protein. Sorghum is rich in carbohydrates and fiber. Moringa greens (leaves) are an excellent source of protein. Dry, powdered leaves indeed are a much-concentrated source of many quality amino acids. Sweet potatoes contain a wealth of orange-hued carotenoid pigments, it has been shown to be a better source of bioavailable beta-carotene than green leafy vegetables, has a high amount of Vitamin A, Vitamin B5, Vitamin B6, Thiamin and Riboflavin. Corn flakes, wheat flakes, and rice flakes are typical examples of flaked cereals. Extruded flakes differ from those made by the traditional process in that the grit for flaking is formed by extruding mixed flour ingredients through a die hole and cutting off pellets of the dough in the desired size [1, 2].

Ready-to-eat breakfast cereals are increasingly gaining acceptance in most developing countries due to their convenience, ease of preparation, and nutritional values. There is an increasing demand for convenient foods and variety as well as nutritional quality and affordability. Nutritional properties and health relevance are a key driving force in flakes manufacturing, the nutritional quality depends both on their composition and structure. The market is further driven by changing food habits, consumers want more transparency on food sourcing, and are increasingly looking for more convenience and healthier options that are instant, high in fiber or protein, low in carbohydrates, and free of artificial colors and flavors. Governments globally are also tightening regulations on nutrition. The flaked cereals business needs to meet consumer expectations in terms of nutrition, health, and taste. Population growth continues and the rate of consumption of instant breakfast is on the increase. New ways will need to be discovered to sustainably grow more breakfast cereals that promotes health, convenient, and meet consumer's nutritional needs [2].

The traditional method used to prepare flaked cereals involves direct cooking of intact grain kernels or parts of kernels with water and flavor in a steam cooker. The basic raw material for the traditionally cooked corn flake comes from the dry milling of regular field corn. The second method involves cooking of finer materials, such as grain flour, in an extruder where mechanical energy is applied for the formation of the grits for flaking [3–7]. There is a current trend of using non-traditional grains, novel ingredients for production of flakes; scientific research into the use of multigrain and fiber is on the increase [8–12]. Unfortunately, practical application in these areas remains proprietary information to each food manufacturer.

Due to changes in lifestyle and urbanization, the consumption of instant flaked cereals has increased in Nigeria. There is the need to formulate value added flakes so as to eliminate the issue of nutrient imbalance in flakes. Formulation of flakes from different grains is one of the ways to improve flasks quality. Hundreds of people have worked on the development of new cereals and the improvement of older ones. There have been new types of raw materials for cereal making introduced over the years [2]. In this study, nutritionally improved, value-added instant flakes were developed, characterized and optimized, via constrained optimal (custom) mixture experimental design, from blends of rice, sorghum, soybean, sweet potato, moringa seed powder and ginger.

2. Materials and methods

2.1 Materials

The major components used for the formulation of the flakes were rice, sorghum and soybean which comprises 60% of the mixture. The other ingredients were water, sugar, malt, egg, sweet potato, ginger and moringa seed powder. All these materials were purchased from Kure Market, Minna, Nigeria.

2.1.1 Processing of the rice, sorghum, and soya beans into flours

Properly clean rice and sorghum grains were milled until fine flour is achieved using a bore mill. Cleaned, sorted soybeans grains were roasted until a golden brown color was observed, and the roasted soybeans were dehulled in a commercial attrition mill, winnowed manually, milled into flour. The flours were sieved using a laboratory sieve mesh of 0.75–1 mm.

2.2 Methods

2.2.1 Experimental design for flakes grits formulation

A three-component constrained optimal (custom) mixture experimental design, with 30 randomized experimental runs, was employed. The formulation design constraints were: rice flour ($30\% \leq x_1 \leq 35\%$), sorghum flour ($20\% \leq x_1 \leq 25\%$), and soybean flour ($5\% \leq x_1 \leq 10\%$). These major components comprise 60% of the total mixture and other ingredients were water (19%), sugar (8%), malt (2%), egg (3%), sweet potato (3%), ginger (2%) and moringa seed powder (3%). The design matrix used for the formulation experiment were presented in **Table 1**. The samples or runs were prepared based on the design matrix. The other minor components were added to each of the 30 samples and mixed together thoroughly, to obtained homogeneous mixture. The samples were then subjected to an electrical steam pressure cooking at temperature of 80°C for 1 hours and then the samples were removed and allowed to cool down for 5 minutes. Each of the samples were then rolled or pressed into flat, thin flakes using a rolling pin, and then were subjected to an electrical oven drying at temperature of 66°C for 1 hours. On removal from the toasting machine, the flakes were allowed to cool down and later packaged in different clean transparent, plastic packaging containers.

	x_1	x_2	x_3	y_{mc}	y_{fat}	y_{ac}	y_{fc}	y_{pc}	y_{nfe}	y_{ev}	y_t	y_f	y_s	y_c	y_{tx}	y_o
Run	%	%	%	%	%	%	%	%	%	k/cal						
1	33.4	20.8	5.8	3.20	3.50	2.50	1.25	29.1	60.5	381	6.8	7.0	7.1	7.4	7.0	7.1
2	33.4	20.8	5.8	3.22	3.51	2.50	1.23	29.0	60.5	381	7.2	7.2	7.3	7.5	7.1	7.3
3	33.4	20.8	5.8	3.20	3.51	2.51	1.25	29.0	60.5	381	7.4	7.3	7.5	7.6	7.5	7.5
4	32.5	20.0	7.5	4.00	2.00	1.50	1.88	30.4	60.2	376	7.8	7.3	7.4	7.2	7.6	7.5
5	32.5	20.0	7.5	4.01	2.02	1.51	1.89	30.5	60.1	376	6.5	6.9	7.0	7.2	6.9	6.9
6	32.5	20.0	7.5	4.00	2.01	1.50	1.87	30.4	60.2	376	7.2	7.1	7.1	7.3	7.1	7.2
7	30.0	25.0	5.0	2.20	3.50	2.50	1.25	29.2	61.3	385	8.0	7.4	7.2	7.5	7.7	7.6
8	30.0	25.0	5.0	2.20	3.51	2.52	1.25	29.2	61.3	385	7.1	7.0	7.3	7.8	7.6	7.4
9	30.0	25.0	5.0	2.21	3.50	2.51	1.26	29.2	61.3	385	6.8	7.1	7.4	7.5	7.1	7.2
10	30.0	22.5	7.5	2.40	4.00	3.00	1.88	28.0	60.7	382	7.0	7.1	7.2	7.6	7.3	7.2
11	30.0	22.5	7.5	2.40	4.01	3.01	1.89	28.0	60.7	382	6.8	7.0	7.1	7.4	7.0	7.1
12	30.0	22.5	7.5	2.42	4.00	3.03	1.88	28.0	60.6	382	7.2	7.2	7.3	7.5	7.1	7.1
13	35.0	20.0	5.0	5.00	3.50	3.50	1.88	28.9	57.2	376	7.4	7.3	7.5	7.6	7.5	7.5
14	35.0	20.0	5.0	5.02	3.50	3.51	1.87	28.9	57.2	376	7.8	7.3	7.4	7.2	7.6	7.5
15	35.0	20.0	5.0	5.01	3.51	3.50	1.88	28.9	57.2	376	6.5	6.9	7.0	7.2	6.9	6.9
16	32.5	22.5	5.0	3.40	3.50	2.50	1.88	26.3	62.5	377	7.2	7.1	7.1	7.3	7.1	7.2
17	32.5	22.5	5.0	3.42	3.51	2.50	1.88	26.2	62.5	377	8.0	7.4	7.2	7.5	7.7	7.6
18	32.5	22.5	5.0	3.41	3.50	2.52	1.87	26.3	62.5	377	7.1	7.0	7.3	7.8	7.6	7.4
19	30.8	20.8	8.4	2.60	4.00	4.00	1.25	17.5	70.7	389	6.8	7.1	7.4	7.5	7.1	7.2
20	30.8	20.8	8.4	2.61	4.00	4.02	1.24	17.5	70.6	389	7.0	7.1	7.2	7.6	7.3	7.2
21	30.8	20.8	8.4	2.60	4.01	4.01	1.25	17.5	70.6	389	6.8	7.0	7.1	7.4	7.0	7.1
22	30.8	23.4	5.8	2.80	3.50	4.00	1.25	26.3	62.2	390	7.2	7.2	7.3	7.5	7.1	7.3
23	30.8	23.4	5.8	2.80	3.53	4.00	1.24	26.3	62.2	390	7.4	7.3	7.5	7.6	7.5	7.5
24	30.8	23.4	5.8	2.82	3.51	4.03	1.25	26.2	62.1	390	7.8	7.3	7.4	7.2	7.6	7.5
25	31.7	21.7	6.6	4.20	3.00	3.00	1.88	29.2	58.7	379	6.5	6.9	7.0	7.2	6.9	6.9
26	31.7	21.7	6.6	4.23	3.01	3.02	1.87	29.2	58.6	379	7.2	7.1	7.1	7.3	7.1	7.2
27	31.7	21.7	6.6	4.22	3.00	3.01	1.88	29.2	58.6	379	8.0	7.4	7.2	7.5	7.7	7.6
28	30.0	20.0	10.0	3.80	2.00	3.00	1.25	328.9	61.1	387	7.1	7.0	7.3	7.8	7.6	7.4
29	30.0	20.0	10.0	3.82	2.01	3.00	1.24	28.9	61.1	387	6.8	7.1	7.4	7.5	7.1	7.2
30	30.0	20.0	10.0	3.81	2.01	3.01	1.25	28.9	61.1	387	7.0	7.1	7.2	7.6	7.3	7.2

x_1 = *Rice flour* (%), x_2 = *Sorghum flour* (%), x_3 = *Soybean flour* (%)
y_{mc} = *Moisture Content* (%); y_{pc} = *Protein Content* (%); y_{fat} = *Fat Content* (%), y_{ac} = *Ash Content* (%)
y_{nfe} = *Nitrogen Free Extract* (%); y_{ev} = *Energy value* (*k/cal*)
y_{fc} = *Fibre Content* (%); y_t = *Taste*; y_f = *Flavor*; y_s = *Sweetness*; y_c = *Colour*; y_{tx} = *Texture*;
y_o = *Overall acceptability*

Table 1.
The design matrix, proximate compositions and the sensory evaluation average scores of the formulated flakes.

2.2.2 Proximate analysis and sensory evaluations

The quality characteristic of the flakes which were determined using the method described by the Association of Analytical Chemist [13] include moisture content, ash content, fat content, crude fiber, crude protein, nitrogen free extract, and energy value. Sensory evaluations of the formulated flakes were also conducted using 30 trained panelists. A 9-point hedonic scale ranging from 9 = like extremely and 1 = dislike extremely was used to evaluate the samples for taste, flavor, sweetness, color, texture and overall acceptability. Table water was used for mouth rinsing intermittently to minimize the carry over effects.

3. Experimental results

The proximate composition of the formulated flakes from rice, sorghum, and soy beans were presented in **Table 1**.

The photos of some of the formulated instant cereal breakfast flakes are presented in **Figure 1**.

Figure 1.
Figure of formulated flakes.

3.1 Statistical analysis of experimental results

The experimental data were analyzed and appropriate Scheffe canonical models were fitted to the mean proximate property data. The statistical significance of the terms in the Scheffe canonical regression models were tested using analysis of variance (ANOVA) for each response, and the adequacy of the models were evaluated by coefficient of determination, F-value, and model p-values at the 5% level of significance. The models were also subjected to lack-of-fit and adequacy tests. The fitted models for each of the response was used to generate 3-D response surface as well as the contour plots using the DESIGN EXPERT 13.0 statistical software.

3.2 Generating the optimal formulation

A numerical optimization approach, exploiting the desirability function technique, was utilized to generate the optimal formulation with the anticipated responses. Optimization goals are assigned to parameters and these goals were used to construct desirability indices (di). Desirability index range from zero to one for any given response and individual desirability for all the responses, in the case of multi-response optimization, are combined into a single number known as overall desirability index. A value of one represents the case where all goals are met perfectly. A zero indicates that one or more responses fall outside the set desirable limits. Under this approach, each *ith* response is assigned a desirability function, d_i, where the value of $d_i$varies between 0 and 1. The function, is defined differently based on the objective of the response.

If the response is to be maximized, then d_i is defined by equation1. If the response is to be minimized, as in the case when the response is cost, is then d_i is defined by Eq. 2. There may be times when the experimenter wants the response to be neither maximized nor minimized, but instead stay as close to a specified target as possible. In such cases, the desirability function is defined by Eq. 3. Once a desirability function is defined for each of the responses, assuming that there are m responses, an overall desirability function is obtained by Eq. 4. The objective is to now find the settings that return the maximum value of D. The rationale for using a geometric rather than an arithmetic mean is that if any individual desirability di is equal to zero, then the overall desirability will also be equal to zero.

$$d_i = \left\{ \begin{array}{cc} 0 & y_i < L \\ \left(\frac{y_i - L}{T - L}\right)^w & L \le y_i \le T \\ 1 & y_i > T \end{array} \right\} \quad (1)$$

where T represents the target value of the *ith* response, y_i, L, represents the acceptable lower limit value for this response and w represents the weight. When $w =$ 1 the function is linear. If $w > 1$ then more importance is placed on achieving the target for the response, y_i. When $w < 1$, less weight is assigned to achieving the target for the response, y_i.

$$d_i = \left\{ \begin{array}{cc} 1 & y_i < T \\ \left(\frac{U - y_i}{U - T}\right)^w & T \le y_i \le U \\ 0 & y_i > U \end{array} \right\} \quad (2)$$

where U represents the acceptable upper limit for the response.

$$d_i = \begin{Bmatrix} 0 & y_i < L \\ \left(\frac{y_i - L}{T - L}\right)^{w_1} & L \le y_i \le T \\ \left(\frac{U - y_i}{U - T}\right)^{w_2} & T \le y_i \le U \\ 0 & y_i > U \end{Bmatrix} \quad (3)$$

$$D = \left(d_1^{r_1}.d_2^{r_2}.d_3^{r_3} \ldots \ldots \ldots d_m^{r_m}\right)^{1/(r_1+r_2+r_3+\ldots\ldots+r_m)} \quad (4)$$

where the r_i's represent the importance of each response. The greater the value of r_i, the more important the response with respect to the other responses.

Numerical optimization solutions are given as a list, in their order of desirability, detailing the components proportions and process variables values that satisfies the set criteria and the overall desirability. The numerical solution can be presented in the form of desirability contour and 3-D Surface plots, optimal bar graph and graphical optimization overlay contour plot; showing the optimized formulation compositions and/or regions that meet specifications [14–17].

4. Results of statistical analysis of experimental data and discussion

The summary of the analysis of variance (ANOVA) for the formulated flake's proximate compositions and the energy value are presented in **Tables 2–8**.

Source	Sum of Mean Squares	df	Square	F-value	p-value	
Model	17.5	8	2.19	11.1	5.22E-06	significant
Linear Mixture	14.4	2	7.22	36.6	1.45E-07	significant
x_1x_2	0.148	1	0.148	0.749	0.397	
x_1x_3	0.487	1	0.487	2.47	0.131	
x_2x_3	0.943	1	0.943	4.78	0.0403	significant
$x_1^2x_2x_3$	0.240	1	0.240	1.22	0.283	
$x_1x_2^2x_3$	2.25	1	2.25	11.4	0.00287	significant
$x_1x_2x_3^2$	0.0835	1	0.0835	0.423	0.523	
Residual	4.15	21	0.197			
Lack of Fit	4.14	4	1.04	9.87E+03	2.31E-28	significant
Pure Error	0.00178	17	0.000105			
Cor Total	21.7	29				
Std. Dev.	0.4443	R^2		0.8087		
Mean	3.3677	Adjusted	R^2	0.7359		
C.V. %	13.19295	Predicted	R^2	0.6566	Adeq Precision	11.5341

Table 2.
ANOVA for moisture content of multigrain flakes.

	Sum of Mean Source	Squares	df	Square	F-value	p-value
Model	13.3	8	1.66	14.2	7.02E-07	significant
Linear Mixture	0.359	2	0.179	1.53	0.240	
x_{12}	0.392	1	0.392	3.35	0.0815	
x_{13}	5.69	1	5.69	48.6	6.98E-07	significant
x_2x_3	0.209	1	0.209	1.78	0.196	
$x_1^2x_2x_3$	1.28	1	1.28	10.9	0.00339	significant
$x_1x_2^2x_3$	1.68	1	1.68	14.3	0.00108	significant
$x_1x_2x_3^2$	2.66	1	2.66	22.7	0.000105	significant
Residual	2.46	21	0.117			
Lack of Fit	2.46	4	0.615	4.90E+03	8.93E-26	significant
Pure Error	0.00213	17	0.000125			
Cor Total	15.7	29				
Std. Dev.	0.3423	R^2	0.8437			
Mean	2.9573	Adjusted R^2	0.7841			
C.V. %	11.5740	Predicted R^2	0.7193	Adeq Precision	10.7149	

Table 3.
ANOVA for fat content of multigrain flakes.

Source	Sum of Mean Squares	df	Square	F-value	p-value	
Model	12.1	8	1.51	18.3	7.84E-08	significant
Linear Mixture	3.99	2	1.99	24.1	3.67E-06	
x_{12}	0.00472	1	0.00472	0.0570	0.814	
x_{13}	0.954	1	0.954	11.5	0.00273	significant
x_2x_3	3.41	1	3.41	41.2	2.30E-06	significant
$x_1^2x_2x_3$	0.00182	1	0.00182	0.0220	0.883	
$x_1x_2^2x_3$	1.58	1	1.58	19.1	0.000269	significant
$x_1x_2x_3^2$	2.22	1	2.22	26.8	3.94E-05	significant
Residual	1.74	21	0.0827			
Lack of Fit	1.74	4	0.434	6.61E+03	7.01E-27	significant
Pure Error	0.00112	17	6.57E-05			
Cor Total	13.8	29				
Std. Dev.	0.2876	R^2	0.8743			
Mean	3.2557	Adjusted R^2	0.8264			
C.V. %	8.8350	Predicted R^2	0.7743	Adeq Precision	13.0502	

Table 4.
ANOVA for ash content of multigrain flakes.

	Sum of Mean					
Source	Squares	df	Square	F-value	p-value	
Model	1.47	8	0.183	2.53	0.0422	significant
Linear Mixture	0.515	2	0.258	3.55	0.0469	significant
x_{12}	0.142	1	0.142	1.96	0.176	
x_{13}	0.142	1	0.142	1.96	0.176	
x_2x_3	0.679	1	0.679	9.36	0.00596	significant
$x_1^2x_2x_3$	0.214	1	0.214	2.95	0.101	
$x_1x_2^2x_3$	0.00529	1	0.00529	0.0730	0.790	
$x_1x_2x_3^2$	0.0108	1	0.0108	0.148	0.704	
Residual	1.52	21	0.0725			
Lack of Fit	1.52	4	0.381	9.03E+03	4.95E-28	significant
Pure Error	0.000717	17	4.22E-05			
Cor Total	2.99	29				
Std. Dev.	0.2693	R^2	0.4908			
Mean	1.563	Adjusted R^2	0.29675			
C.V. %	17.2298	Predicted R^2	0.0855	Adeq Precision	4.2265	

Table 5.
ANOVA for crude fiber of multigrain flakes.

Source	**Sum of Mean Squares**	**df**	**Square**	**F-value**	**p-value**	
Model	276	8	34.5	7.56	9.49E-05	significant
Linear Mixture	24.1	2	12.1	2.65	0.0943	
x_{12}	19.8	1	19.8	4.35	0.0494	significant
x_{13}	2.52	1	2.52	0.553	0.465	
x_2x_3	4.28	1	4.28	0.939	0.344	
$x_1^2x_2x_3$	63.3	1	63.3	13.9	0.00125	significant
$x_1x_2^2x_3$	17.9	1	17.9	3.92	0.0610	
$x_1x_2x_3^2$	204	1	204	44.8	1.28E-06	significant
Residual	95.7		21	4.56		
Lack of Fit	95.7	4	23.9	2.03E+04	5.00E-31	significant
Pure Error 0.0200	17		0.00118			
Cor Total	371		29			
Std. Dev.	2.1346		R^2	0.7423		
Mean	27.37		Adjusted R^2	0.6442		
C.V. %	7.7989		Predicted R^2	0.5375	Adeq Precision	9.9719

Table 6.
ANOVA for crude protein of multigrain flakes.

Source	Sum of Mean Squares	df	Square	F-value	p-value	
Model	241	8	30.2	6.07	0.000426	significant
Linear Mixture	66.7	2	33.4	6.71	0.00558	significant
x_{12}	26.4	1	26.4	5.31	0.0316	significant
x_{13}	4.21	1	4.21	0.847	0.368	
x_2x_3	0.0221	1	0.0221	0.00445	0.947	
$x_1^2x_2x_3$	34.6	1	34.6	6.96	0.0154	significant
$x_1x_2^2x_3$	32.3	1	32.3	6.50	0.0187	significant
$x_1x_2x_3^2$	133	1	133	26.7	4.04E-05	significant
Residual	104	21	4.97			
Lack of Fit	104	4	26.1	1.40E+04	1.19E-29	significant
Pure Error	0.0317	17	0.00186			
Cor Total	346	29				
Std. Dev.	2.2294	R^2	0.6981			
Mean	61.4867	Adjusted R^2	0.5831			
C.V. %	3.6259	Predicted R^2	0.4580	Adeq Precision	8.6721	

Table 7.
ANOVA for nitrogen free extract of multigrain flakes.

Source	Sum of Mean Squares	df	Square	F-value	p-value	
Model	488	6	81.3	6.85	0.000288	significant
Linear Mixture	387	2	194	16.3	3.86E-05	significant
x_{12}	18.1	1	18.1	1.52	0.230	
x_{13}	51.9	1	51.9	4.37	0.0478	significant
x_2x_3	18.6	1	18.6	1.57	0.223	
$x_1x_2x_3$	85.6	1	85.6	7.22	0.0132	significant
Residual	273	23	11.9			
Lack of Fit	273	6	45.5			
Pure Error	0.000	17	0.000			
Cor Total	761	29				
Std. Dev.	3.4447	R^2	0.6413			
Mean	382.2	Adjusted R^2	0.5477			
C.V. %	0.9013	Predicted R^2	0.4912	Adeq Precision	7.0182	

Table 8.
ANOVA for Energy Value of multigrain flakes.

The moisture content fitted model in terms of L_Pseudo Components is presented in Eq. (5):

$$\left.\begin{aligned} y_{mc} = {} & 4.92x_1 + 2.11x_2 + 3.73x_3 - 1.09x_1x_2 - 1.97x_1x_3 \\ & -2.74x_2x_3 - 29.2x_1^2x_2x_3 + 89.3x_1x_2^2x_3 - 17.2x_1x_2x_3^2 \end{aligned}\right\} \tag{5}$$

The results of the analysis showed that the moisture content model of the formulated instant flakes is significant with F-value of 11.1 and p-value of 5.22E-06. The moisture content is significantly influenced, at 5% level of significance, by the proportions of rice, sorghum, and soybean flours in the formulations (with linear mixture F- and p-values of 36.6 and 1.45E-07, respectively). The moisture content is also significantly influenced, at 5% level of significance by the sorghum/soybean flours interaction (with F-value of 4.78 and p-value of 0.0403); and rice/the second order of sorghum/soybean flours interaction (with F-value of 11.4 and p-value of 0.00287). The Lack of Fit F-and p-value of 9.87E+03 and 2.31E-28 implies that the Lack of Fit is significant. The moisture content model R^2 and the Adjusted R^2 are 0.8087 and 0.7359, respectively. The predicted R^2 of 0.6566 is in reasonable agreement with the adjusted R^2 of 0.7359; i.e. the difference is less than 0.2. Adequacy of precision ratio of 11.534 indicates an adequate signal. This model can be used to navigate the design space and can be used to make predictions about moisture content for given levels of each factor.

The fat content fitted model in terms of L_Pseudo Components is presented in Eq. (6):

$$\left.\begin{aligned} y_{fat} = {} & 3.57x_1 + 2.58x_2 + 3.06x_3 - 1.77x_1x_2 - 6.73x_1x_3 \\ & +1.29x_2x_3 - 67.4x_1^2x_2x_3 + 77.2x_1x_2^2x_3 + 97.0x_1x_2x_3^2 \end{aligned}\right\} \tag{6}$$

The results of the analysis showed that the fat content model of the formulated instant flakes is significant with F-value of 14.2 and p-value of 7.02E-07. The fat content is not significantly influenced, at 5% level of significance, by the proportions of rice, sorghum, and soybean flours in the formulations (with linear mixture F- and p-values of 1.53 and 0.240, respectively). The fat content is significantly influenced, at 5% level of significance by the rice/soybean flours interaction (with F- value of 48.6 and p-value of 6.98E-07); the second order of rice/sorghum/soybean flours interaction (with F-value of 10.9 and p-value of 0.00339); rice/the second order of sorghum/soybean flours interaction (with F-value of 14.3 and p-value of 0.00108), and rice/sorghum/the second order of soybean flours interaction (with F-value of 22.7 and p-value of 0.000105). The Lack of Fit F-and p-value of 4.90E+03 and 8.93E-26 implies that the Lack of Fit is significant. The fat content model R^2 and the Adjusted R^2 are 0.8437 and 0.7841, respectively. The predicted R^2 of 0.7193 is in reasonable agreement with the adjusted R^2 of 0.7841; i.e. the difference is less than 0.2. Adequacy of precision ratio of 10.715 indicates an adequate signal. This model can be used to navigate the design space and can be used to make predictions about the fat content for given levels of each factor.

The ash content fitted model in terms of L_Pseudo Components is presented in eq. (7):

$$\left.\begin{aligned} y_{ac} = {} & 3.56x_1 + +3.56x_2 + 2.06x_3 + 0.194x_1x_2 - 2.76x_1x_3 \\ & +5.21x_2x_3 - 2.54x_1^2x_2x_3 - 74.9x_1x_2^2x_3 + 88.7x_1x_2x_3^2 \end{aligned}\right\} \tag{7}$$

The results of the analysis showed that the ash content model of the formulated instant flakes is significant with F-value of 18.3 and p-value of 7.84E-08. The ash content is significantly influenced, at 5% level of significance, by the proportions of rice, sorghum, and soybean flours in the formulations (with linear mixture F- and p-values of 24.1 and 3.67E-06, respectively). The ash content is also significantly influenced, at 5% level of significance by the rice/soybean flours interaction (with F-value of 11.5 and p-value of 0.00273); sorghum/soybean flours interaction (with F-value of 41.2 and p-value of 2.30E-06); rice/the second order of sorghum/soybean flours interaction (with F-value of 19.1 and p-value of 0.000269); and rice/sorghum/the second order of soybean flours interaction (with F-value of 26.8 and p-value of 3.94E-05). The Lack of Fit F-and p-value of 6.61E+03 and 7.01E-27 implies that the Lack of Fit is significant. The ash content model R^2 and the Adjusted R^2 are 0.8743 and 0.8264, respectively. The predicted R^2 of 0.7743 is in reasonable agreement with the adjusted R^2 of 0.8264; i.e. the difference is less than 0.2. Adequacy of precision ratio of 13.05 indicates an adequate signal. This model can be used to navigate the design space and can be used to make predictions about ash content for given levels of each factor.

The crude fiber fitted model in terms of L_Pseudo Components is presented in eq. (8):

$$\left.\begin{aligned} y_{fc} &= 1.82x_1 + 1.20x_2 + 1.20x_3 + 1.07x_1x_2 + 1.06x_1x_3 \\ &\quad +2.33x_2x_3 - 27.6x_1^2x_2x_3 - 4.33x_1x_2^2x_3 - 6.17x_1x_2x_3^2 \end{aligned}\right\} \quad (8)$$

The results of the analysis showed that the crude fiber model of the formulated instant flakes is significant with F-value of 2.53 and p-value of 0.0422. The crude fiber is significantly influenced, at 5% level of significance, by the proportions of rice, sorghum, and soybean flours in the formulations (with linear mixture F- and p-values of 3.55 and 0.0469, respectively). The crude fiber is also significantly influenced, at 5% level of significance by sorghum/soybean flours interaction (with F-value of 9.36 and p-value of 0.00596); The Lack of Fit F- and p-value of 9.03E+03 and 4.95E-28 implies that the Lack of Fit is significant. The crude fiber model R^2 and the Adjusted R^2 are 0.4908 and 0.29675, respectively. The predicted R^2 of 0.0855 is not close to the adjusted R^2 of 0.2967; i.e., the difference is more than 0.2. This indicates a possible problem with the fitted model. Adequacy of precision ratio of 4.227 still indicates an adequate signal. Thus, the model can still be used to navigate the design space and to make predictions about crude fiber for given levels of each factor.

The crude protein fitted model in terms of L_Pseudo Components is presented in Eq. (9):

$$\left.\begin{aligned} y_{pc} &= 28.5x_1 + 28.8x_2 + 28.5x_3 - 12.6x_1x_2 + 4.48x_1x_3 \\ &\quad -5.84x_2x_3 + 474.x_1^2x_2x_3 + 252.x_1x_2^2x_3 - 850.x_1x_2x_3^2 \end{aligned}\right\} \quad (9)$$

The results of the analysis showed that the crude protein model of the formulated instant flakes is significant with F-value of 7.56 and p-value of 9.49E-05. The crude protein is not significantly influenced, at 5% level of significance, by the proportions of rice, sorghum, and soybean flours in the formulations (with linear mixture F- and p-values of 2.65 and 0.0943, respectively). The crude protein is significantly influenced, at 5% level of significance by rice/sorghum flours interaction (with F- value of 4.35 and p-value of 0.0494); second order of rice/sorghum/soybean flours interaction (with F-value of 13.9 and p-value of 0.00125); and rice/sorghum/the second order of soybean flours interaction (with F-value of 44.8 and p-value of 1.28E-06). The Lack of Fit F-and p-value of 2.03E+04 and 5.00E-31 implies that the Lack of Fit is significant. The crude protein model R^2 and the Adjusted R^2 are 0.7423 and 0.6442, respectively. The predicted R^2 of 0.5375 is in reasonable agreement with the adjusted R^2 of 0.6442; i.e. the difference is less than 0.2. Adequacy of precision ratio of 9.972 indicates an adequate signal. This model can be used to navigate the design and can be used to make predictions about the crude protein for given levels of each factor.

The nitrogen free extract fitted model in terms of L_Pseudo Components is presented in Eq. (10):

$$\left.\begin{aligned} y_{nfe} &= 57.6x_1 + 61.7x_2 + 61.5x_3 + 14.5x_1x_2 + 5.79x_1x_3 \\ &\quad -0.42x_2x_3 - 351.x_1^2x_2x_3 - 339.x_1x_2^2x_3 + 686.x_1x_2x_3^2 \end{aligned}\right\} \tag{10}$$

The results of the analysis showed that the nitrogen free extract model of the formulated instant flakes is significant with F-value of 6.07 and p-value of 0.000426. The nitrogen free extract is significantly influenced, at 5% level of significance, by the proportions of rice, sorghum, and soybean flours in the formulations (with linear mixture F- and p-values of 6.71 and 0.00558, respectively). The nitrogen free extract is also significantly influenced, at 5% level of significance by rice/sorghum flours interaction (with F-value of 5.31 and p-value of 0.0316); second order of rice/sorghum/soybean flours interaction (with F-value of 6.96 and p-value of 0.0154); rice/second order of sorghum/soybean flours interaction (with F-value of 6.50 and p-value of 0.0187); and rice/sorghum/the second order of soybean flours interaction (with F-value of 26.7 and p-value of 4.04E-05). The Lack of Fit F-and p-value of 1.40E+04 and 1.19E-29 implies that the Lack of Fit is significant. The nitrogen free extract model R^2 and the Adjusted R^2 are 0.6981 and 0.5831, respectively. The predicted R^2 of 0.4580 is in reasonable agreement with the adjusted R^2 of 0.5831; i.e. the difference is less than 0.2. Adequacy of precision ratio of 8.672 indicates an adequate signal. This model can be used to navigate the design space and can be used to make predictions about nitrogen free extract for given levels of each factor.

The energy value fitted model in terms of L_Pseudo Components is presented in Eq. (11):

$$\left.\begin{aligned} y_{ev} &= 376x_1 + 386.x_2 + 388.x_3 - 11.9x_1x_2 - 20.2x_1x_3 \\ &\quad -12.1x_2x_3 + 172x_1x_2x_3 \end{aligned}\right\} \tag{11}$$

The results of the analysis showed that the energy value model of the formulated instant flakes is significant with F-value of 6.85 and p-value of 0.000288. The energy value is significantly influenced, at 5% level of significance, by the proportions of rice, sorghum, and soybean flours in the formulations (with linear mixture F- and p-values of 16.3 and 3.86E-05, respectively). The energy value is also significantly influenced, at 5% level of significance by rice/soybean flours interaction (with F-value of 4.37 and p-value of 0.0478); and rice/sorghum/soybean flours interaction (with F-value of 7.22 and p-value of 0.0132). The energy value model R^2 and the Adjusted R^2 are 0.6413 and 0.5477, respectively. The predicted R^2 of 0.4912 is in reasonable agreement with the adjusted R^2 of 0.5477; i.e. the difference is less than 0.2. Adequacy of precision ratio of 7.018 indicates an adequate signal. This model can be used to navigate the design space and can be used to make predictions about energy value for given levels of each factor.

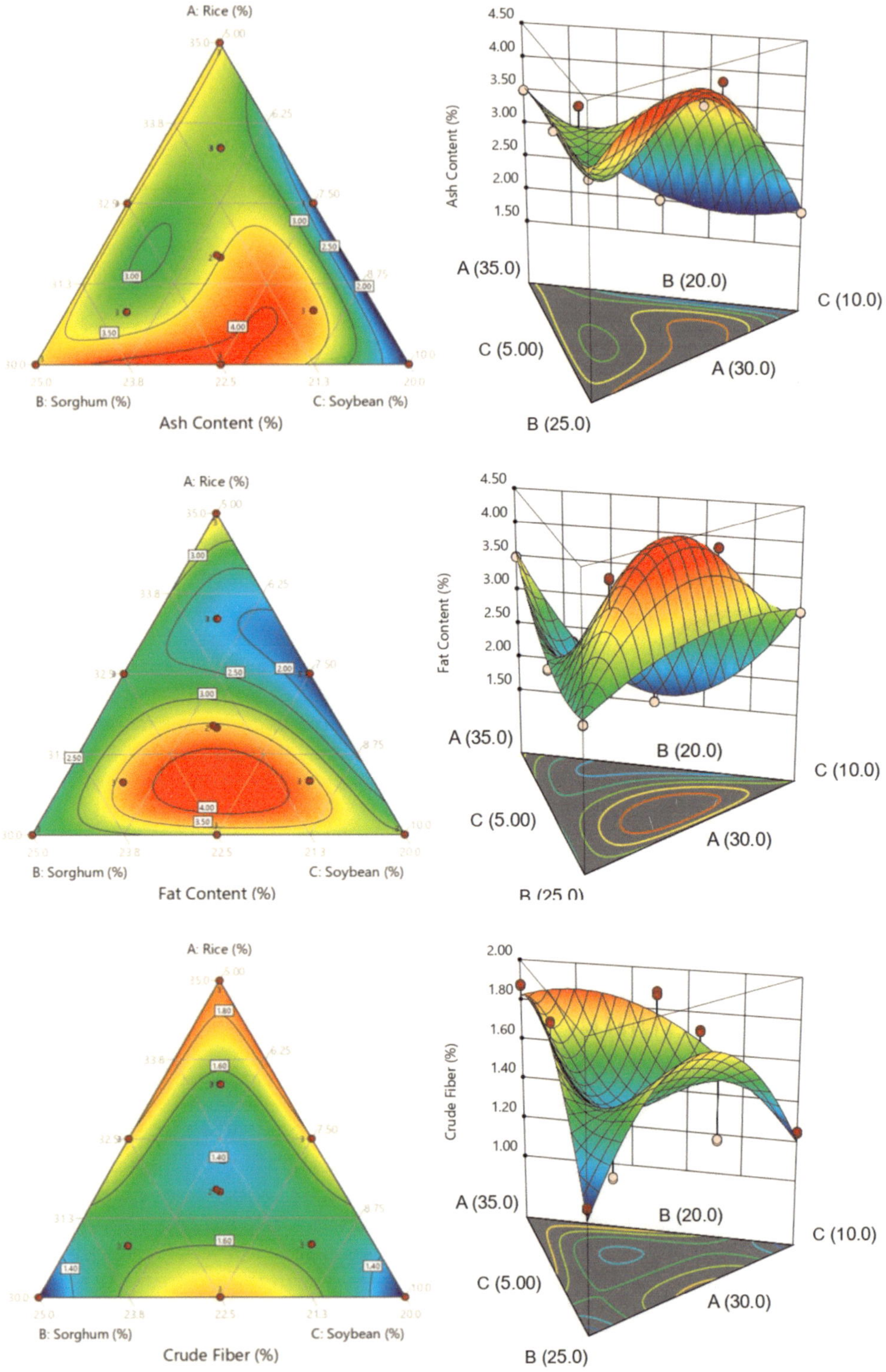
A: Rice (%)
B: Sorghum (%)
C: Soybean (%)
Ash Content (%)
4.50
4.00
3.50
3.00
2.50
2.00
1.50
Ash Content (%)
A (35.0)
B (20.0)
C (10.0)
C (5.00)
A (30.0)
B (25.0)
A: Rice (%)
B: Sorghum (%)
C: Soybean (%)
Fat Content (%)
4.50
4.00
3.50
3.00
2.50
2.00
1.50
Fat Content (%)
A (35.0)
B (20.0)
C (10.0)
C (5.00)
A (30.0)
B (25.0)
A: Rice (%)
B: Sorghum (%)
C: Soybean (%)
Crude Fiber (%)
2.00
1.80
1.60
1.40
1.20
1.00
Crude Fiber (%)
A (35.0)
B (20.0)
C (10.0)
C (5.00)
A (30.0)
B (25.0)

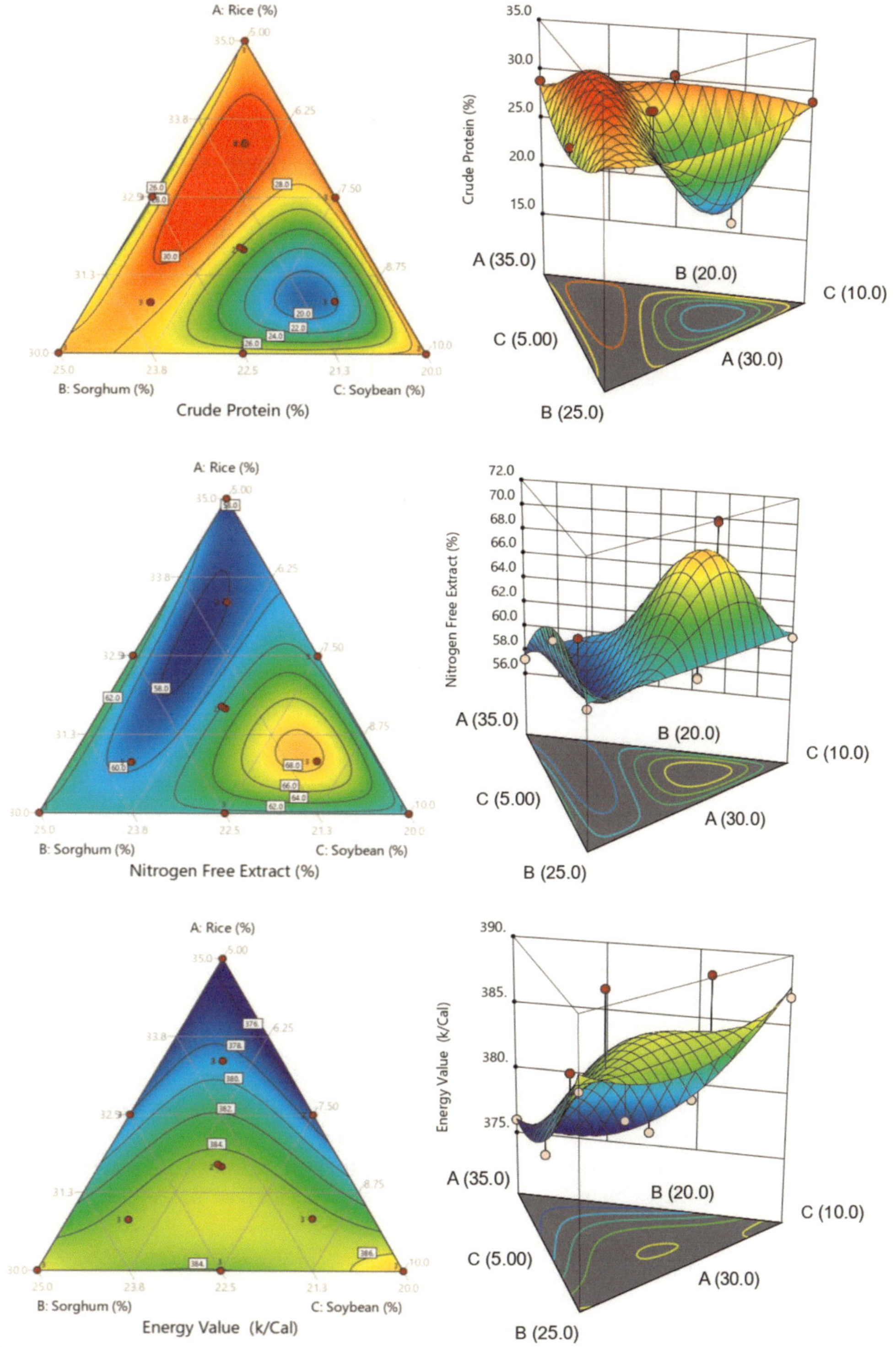

Figure 2.
The contours and 3-D plots for the proximate compositions, nitrogen free extract, and energy value of multigrain flakes.

Name	Goal	Lower Limit	Upper Limit	Lower Weight	Upper Weight	Importance
Rice	in range	30	35	1	1	3
Sorghum	in range	20	25	1	1	3
Soybean	target = 10	5	10	1	10	5
Moisture Content	target = 2.5	2.2	5	1	1	3
Ash Content	target = 3	2	4	1	1	3
Fat Content	in range	1.5	4	1	1	3
Crude Fiber	target = 1.8	1.23	1.89	1	1	3
Crude Protein	target = 30	20	30.5	1	10	5
Nitrogen Free Extract	minimize	57.2	60	1	1	3
Energy Value	target = 390	376	390	1	10	5
Taste	in range	6.5	8	1	1	3
Flavor	in range	6.9	7.4	1	1	3
Texture	in range	7	7.5	1	1	3
Color	in range	7.2	7.8	1	1	3
Sweetness	in range	6.9	7.7	1	1	3
Overall Acceptability	maximize	6.9	7.56	1	1	3

Table 9.
Optimization constraints for instant flakes formulation.

The contours and 3-D plots for the proximate compositions (moisture content, fat content, ash content, crude fiber, crude protein, nitrogen free extract, and energy value) are summarized in **Figure 2**.

Table 9 presents the summary of the optimization constraints employed in the optimization module. The five desirability solutions that were found are presented in **Table 10**. The numerical solution desirability contour plot and 3-D Surface were presented in **Figure 3**. The numerical solution, presented in the form of optimal flake's bar graph and the graphical optimization overlay contour plot, showing the optimized formulation compositions with the respective quality parameters, are summarized in **Figure 4**.

The result of the flakes optimization gave optimized multigrain instant flakes with overall desirability index of 0.519, based on the set optimization goals and individual quality desirability indices. Formulating instant flake with 31.9% rice flour, 22% sorghum flour, 6.05% soybean flour yielded an improved instant flake with optimal quality properties.

5. Conclusions

Instant flakes were developed, characterized and optimized, via constrained optimal (custom) mixture experimental design, from blends of rice, sorghum and soybean. Some concluding observations from the investigation are given below.

No	y_{mc}	y_{ac}	y_{fat}	y_{fc}	y_{pc}	y_{nfe}	y_{ev}	y_{ta}	y_{flav}	y_{tx}	y_{cl}	y_{sw}	y_{oa}	D_i	
1	**3.668**	**3.083**	**3.178**	**1.441**	**30.000**	**58.611**	**383.678**	**7.238**	**7.157**	**7.500**	**7.435**	**7.288**	**7.284**	**0.519**	**Selected**
2	3.493	3.142	3.781	1.552	28.486	59.539	384.864	7.265	7.166	7.500	7.484	7.307	7.309	0.483	
3	3.474	3.256	2.483	1.413	30.000	59.336	381.965	7.206	7.147	7.000	7.377	7.249	7.263	0.463	
4	3.543	3.132	2.686	1.602	29.999	59.071	380.513	7.303	7.177	7.500	7.498	7.401	7.306	0.404	
5	4.224	3.278	2.587	1.738	30.000	58.169	376.628	7.257	7.162	7.000	7.364	7.295	7.261	0.305	
6	4.082	3.239	2.413	1.695	30.500	58.064	377.048	7.250	7.160	7.372	7.363	7.290	7.261	0.035	

y_{mc} = *Moisture Content* (%), y_{pc} = *Protein Content* (%), y_{fat} = *Fat Content* (%), y_{ac} = *Ash Content* (%), y_{nfe} = *Nitrogen Free Extract*(%), y_{ev} = *Energy value*, y_{fc} = *Fibre Content* (%), y_{ta} = *Taste*, y_{flav} = *Flavour*, y_{sw} = *Sweetness*, y_{cl} = *Colour*, y_{tx} = *Texture*, y_{oa} = *Overall Acceptability*, D_i = *Overall Desirability*

Table 10.
The desirability solutions found.

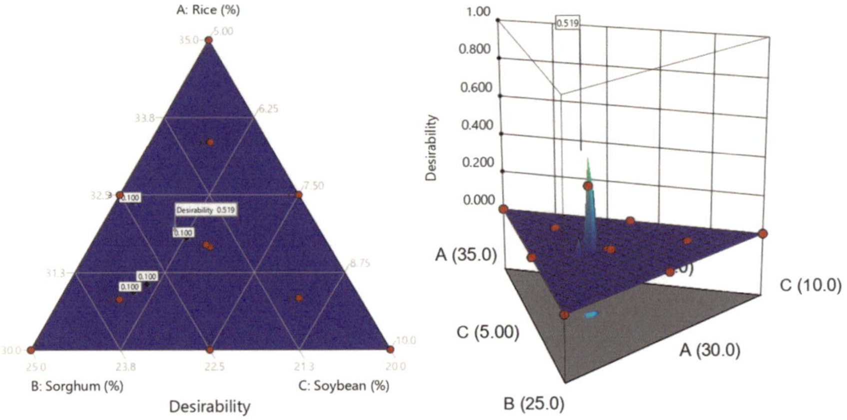

Figure 3.
The numerical solution desirability contour plot and 3-D surface.

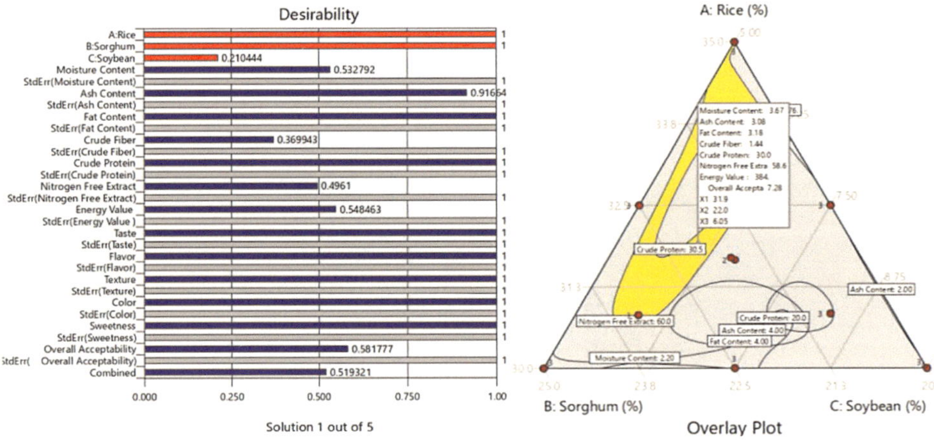

Figure 4.
The bar graph and the graphical optimization overlay contour plot for the optimal formulated flake.

- The optimal flake was obtained from 31.9% rice flour, 22% sorghum flour, 6.05% soybean flour.

- The quality properties of the optimal flake were: 3.67% moisture content, 3.18% fat content, 3.08% ash content, 1.44% crude fiber, 30.0% crude protein, 58.6% nitrogen free extract, 384 kcal energy value and 7.28 overall acceptability

- A The optimal flake from blends of rice flour, sorghum flour, and soybean meal has high nutritional qualities suitable for improving and solving the problems of malnutrition especially in the African continent.

The research has shown that composite food formulation is an excellent way to achieve nutrition revolution, the road to healthier diets and optimal nutrition in Africa: The continent is blessed with vast varieties of agricultural produce seasonally (tubers, roots, cereals, pulses, fruits, vegetables, etc.), yet hunger, malnutrition,

DOI: http://dx.doi.org/10.5772/intechopen.109820

dietary deficit, concurrent diseases and food insecurity persists. Additive food manufacturing and/or composite food formulation, dietary diversification, food fortification and increasing access to varieties of nutritionally adequate foods are vital strategies to tackle these lingering problems. However, this study encouraged that further study be carried out on formulation of instant flakes using other nutritionally rich blends (grains and legumes).

Declaration of interest

The authors declare no conflicts of interest. The authors alone are responsible for the content and writing of the manuscript.

Funding source

This research is self-sponsored and did not receive any specific grant from funding agencies in the public, commercial, or not-for-profit sectors.

Author details

Samuel Tunde Olorunsogo* and Bolanle Adenike Adejumo
Department of Agricultural and Bioresources Engineering, School of Infrastructure, Process Engineering and Technology, Federal University of Technology, Minna, Nigeria

*Address all correspondence to: solorunsogo@futminna.edu.ng

References

[1] Mbaeyi LE, Onweluzo. Effect of sprouting and pregelatinization on the composition and sensory properties of flaked breakfast cereals produced from Sorghum Pigeon Pea Blend. Agro-Science Journal of Tropical Agriculture, Food Environment and Extention. 2010; **9**(3):184-192

[2] Alicia AP, Schonauer SL, Poutanen KS, editors. Breakfast Cereals and How They Are Made - Raw Materials, Processing, and Production. Third ed. Cereals & Grains Association. United Kingdom: Woodhead Publishing, an imprint of Elsevier. 2020. pp. 6-11

[3] Elke KA, Emanuele Z. Cereal grains for the food and beverage industries. In: Woodhead Publishing Series in Food Science, Technology and Nutrition: Number 248. Cambridge UK: Woodhead Publishing Limited; 2013. pp. 91-100

[4] Desalegn G. Proximate composition of complementary food prepared from maize (Zea mays), soybean (Glycine max) and Moringa leaves in Tigray, Ethiopia. Cogent Food & Agriculture. 2019;**5**:1627779. DOI: 10.1080/23311932.2019.1627779

[5] Roberson JA, Crowe CT. This is How to Cite a Book. 6th ed. New York: John Wiley and Sons Inc.; 1997

[6] Sudarshan RN, Janitha PDW, Laurie S. Sustainable Protein Sources. London, United Kingdom: Academic Press, an imprint of Elsevier; 2017. p. 39, 40, 75

[7] Krasnov NF. In: Leib G, editor. Another Example of Book Reference. Translated from the Russian language by. Moscow: MIR Publisher; 1985

[8] Hettiarachchy NS, Ju ZY, Siebenmorgen T, Sharp RN. Rice: Production, processing and utilization. In: Kulp K, Ponte JG, editors. Handbook of Cereal Science and Technology. New York, NY: Marcel Dekker, Inc.; 2000. pp. 203-221

[9] Delcour J, Hoseney RC. Principles of Food Science and Technology. St. Paul, MN, USA: AACC International. Inc.; 2010

[10] Okafor GI, Usman GO. Production and evaluation of breakfast cereals from blends of African yam bean (*Sphenostylis stenocarpa*), maize (zea mays) and defatted coconut (cocus nucifera). Journal of Food Processing and Preservation. 2014;**38**(2014): 1037-1043

[11] Arendt EK, Zannini E. Cereal Grains for the Food and Beverage Industry. Cambridge, UK: Woodhead Publishing Ltd; 2013

[12] Sawant AA, Thakor NJ, Swami SB, Divate AD. Physical and sensory characteristics of Ready-To-Eat food prepared from finger millet based composite mixer by extrusion. Agricultural Engineering International: The CIGR e-journal. 2013;**15**(1):100-105

[13] AOAC. Official Methods of Analysis. 30th ed. Washington: Association of Analytical Chemists; 2002

[14] Smith WF. Experimental Design for Formulation. The ASA-SIAM Series on Statistics and Applied Probability is published jointly by the American Statistical Association and the Society for Industrial and Applied Mathematics. United States of America: Society for Industrial and Applied Mathematics; 2005. pp. 281-290

[15] ReliaSoft. Experiment Design and Analysis Reference. USA. Available

from: http://www.ReliaSoft.com: ReliaSoft Corporation Worldwide Headquarters; 2015. pp. 205-209

[16] Myers RH, Montgomery DC, Anderson-Cook CM. Response surface methodology - Process and product optimization using designed experiments. In: Wiley Series in Probability and Statistics. Hoboken, New Jersey: John Wiley & Sons, Inc.; 2016. pp. 341-343

[17] Selvamuthu D, Das D. Introduction to Statistical Methods - Design of Experiments and Statistical Quality Control. Singapore: Springer Nature Singapore Pte Ltd; 2018. pp. 344-346

Section 3

Food Processing in Food Industry

Chapter 5

Recent Developments in Processing Technologies for Roasted, Fried, Smoked and Fermented Food Products

Olugbenga Ben Ogunmoyela and Musa Omotayo Jimoh

Abstract

Progress in the research efforts to upgrade various traditional processing technologies, especially roasting, deep-fat frying, smoking, and fermentation, is presented in this chapter. The importance of these studies is in the need for more user-friendly, adaptable, and affordable low-cost machinery and equipment for sustainable food processing, especially in communities where electricity is a challenge, and alternative energy sources such as gas and charcoal are critical. The design considerations and characteristics of the various types of machinery as well as the design calculations and performance evaluation results aimed at standardizing the upgraded machinery are therefore presented from various studies. The effects of these technologies on the quality of the resulting products are discussed particularly in relation to the variations in process losses of micronutrients in the fortified products, with examples of vitamin A and iron losses in pan bread and fried doughnuts obtained from baking fortified wheat flours.

Keywords: roasting, frying, smoking, fermentation, product quality, fortification

1. Introduction

Recent progress in the research efforts to upgrade four traditional processing technologies of roasting, frying, smoking, and fermented products is examined in this chapter. Most food processing activities in rural and urban areas in developing countries, whether at the household level or especially at the micro- and small-scale business levels, have for centuries depended on the use of these technologies at the rural scale for human existence [1, 2]. These upgrade efforts are therefore aimed at making life less burdensome by developing more user-friendly and affordable low-cost machinery, in recognition of changing socioeconomic circumstances [3]. This interest has continued to engage the focus of agricultural engineers and food scientists in the last two decades, especially with the prevalence of electricity challenges in most rural communities all over the world and the need for alternative sources of energy.

However, many research studies reported in the literature had until recently been aimed at standardizing the processing technologies for traditional crops like cassava and plantain into dry flour forms to extend their shelf-life [2, 4, 5]. However, in recent years, work has shifted to standardizing the engineering parameters applicable to each of these processing technologies through a better understanding of the fundamentals of the design characteristics, design calculations, and performance evaluation, to enable the construction of such upgraded devices for markets. Testing and validation results of the functionality of these upgrades have also shown the tremendous advances that have been made in the achievement of these objectives [6–8].

The effects of these technologies on the quality of fortified products, in the face of the increasing role of fortification in addressing micronutrient deficiencies in many developing countries, are also examined in this chapter. The likely process losses associated with roasted, fried, baked, and fermented products are discussed, and examples of the associated process losses of baked fortified products, namely pan bread and fried doughnuts, are presented. Losses in vitamin A and iron in these products are shown to be as high as 30–40%, which shows the importance of the type of vitamin-mineral premixes suitable for use in different products and the relationship with the type of processing technology to be adopted.

1.1 Roasting

In roasting, various agricultural products such as yams, plantains, maize, and potatoes are exposed to dry heat over a fire, or in an oven, and cooked to a state in which sufficient moisture has been removed to make the products suitable for immediate consumption. For example, the traditional method of roasting plantain is to place a metal grill on top of a metal pot containing burning charcoal, which supplies the heat to roast the plantain. The use of uncontrolled heat sources [9] has been the most common application of this technology, with an emphasis on convective heat transfer, rather than on a more efficient process by a combination of conduction, convection, and radiation. However, it is noteworthy that recent research reports have examined the application of emerging and more novel techniques such as infra-red heating to the roasting of nuts. The limitations of these technologies have also been highlighted [10, 11].

The increasing demand for roasted products such as semi-ripe plantains has brought about recent interest to upgrade the local method of roasting to be more user-friendly, hygienic, and versatile. In fact, studies on the development of a low-cost and affordable multi-heat source plantain roaster have been undertaken and the performance characteristics of such roaster examined from the effects of using different heat sources, namely, charcoal, gas, and electricity, on the proximate and micronutrient composition of the roasted product. The roasting process has been reported to prevent toxic hydrocarbons as well as microbial contamination. One other advantage reported is the suitability of such products for diabetics [12, 13].

Many studies have also been reported on the development of roasting machines [6, 7, 9]. During the roasting process, there is a transmission of energy from the heat source to the plantain as a result of a temperature gradient. This alters the quality of the food with an enhanced flavor due to some desirable physical and chemical changes. This desirable quality has been attributed to the reduction of the water activity at the surface of the food [14]. Whether the source of heat is gas, electricity, or charcoal, the use of high temperature in roasting facilitates complex changes in the components of the food at the surface and retention of moisture in the interior of the food product. A specific example of such an upgraded and versatile technology is given here, which has high commercial potential in many rural communities as it can employ either charcoal, gas, or electricity, as heat sources for the roasting of plantains, traditionally known as "*Booli*" [15].

1.1.1 Methodology

1.1.1.1 Design characteristics of the multi-heat source plantain roaster

The plantain roaster was designed to use any of the three alternative energy sources, namely, electricity, gas, and standard charcoal. The major components (**Figure 1**) have been described in detail [14] and comprise a cylindrical stainless steel basket, a thermostat, heater, and fan, among others. Separate sections are provided for a gas burner, electric heater, and charcoal departments, while the roasting net is held by a stainless steel standing support and designed in such a way as to enable free movement of the turning handle, which fits into a fabricated frame. Free movement of the heat by conduction, convection, and radiation was found possible with this arrangement [14]. The electric heater is welded to the frame, fitted with an electric heater with a power of 1800 W and a sensor to detect changes in temperature and trip off automatically at the set temperature. The design also holds a load of 2 kg charcoal for heating the plantain in the compartment provided, while the gas cylinder is connected to the burner with a long hose (**Figure 1**).

1.1.1.2 Performance Trials

The plantain roaster was observed to be heated to 200°C while empty with both electric and gas sources and to 190°C when using charcoal as the heat source. Temperature regulation is achieved for the electric heater with an electric switch that regularly trips off and switches back on as necessary. The cycle of pressure decrease

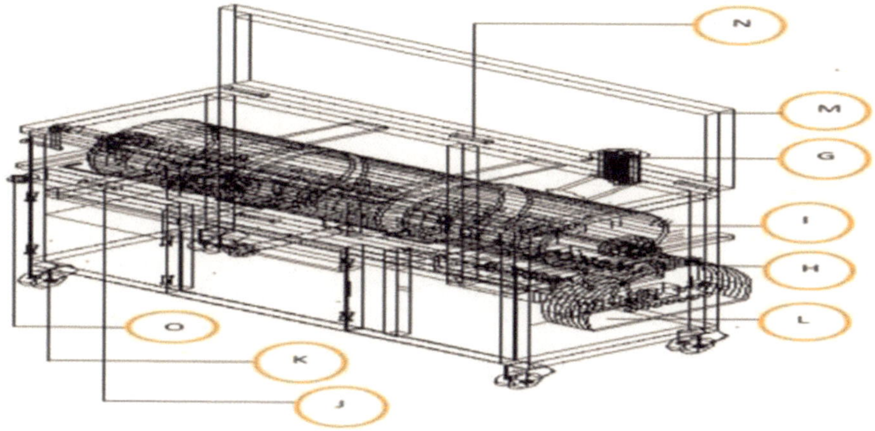

Figure 1.
The design of the plantain roaster: G—chimney, H—the turning handle, I—roasting basket, J—door handle, K—the tyre, L—fan compartment, M—lid cover, N—hinges, O—control switch.

and increase is used to control the rate at which the gas flows, with gas regulated on the basis of the color of the flame produced, while the charcoal is tested by the level of heat it produces when empty. The efficiency of the charcoal when it produces energy is tested compared to the volume of charcoal in the compartment [14].

The optimum temperature for both electric and gas sources was observed to be 200°C while that of charcoal was 190°C, these being the temperatures at which the plantain attained an acceptable color in the shortest time. The ambient temperature was 30°C.

1.1.1.3 Performance calculations

$$\text{Using the eq.} Q = -KA\frac{dt}{dx} \tag{1}$$

where Q is heat transferred by conduction, K is the thermal conductivity of the stainless steel roasting net, and the minus sign indicates that it obeys the second law of thermodynamics, A is the cross-sectional area of the roasting net through which heat flows to the plantain, and $\partial T/\partial x$ is the temperature gradient in the direction of heat flow.

$$\mathrm{Q} = \mathrm{Q_{conv}} + \mathrm{Q_{cond}} + \mathrm{Q_{rad}} \tag{2}$$

$$\mathrm{Q} = \left[hA\left(t_s - t_f\right) + \left(-K_pA\frac{dt}{dx}\right)\right] + \mathrm{Q_{rad}} \tag{3}$$

where Q_{conv}, Q_{cond}, and Q_{rad} represent heat transfer by convection, conduction, and radiation, respectively; h is the convective heat transfer coefficient of 31.77 W/m^2°C [16]; K_p is the thermal conductivity of the plantain 0.594 W/m°C; t_f is average room temperature before heating in °C; and t_s is surface temperature of the plantain in °C.

The mass (m_1) of different unripe plantains was weighed to an average of 130 g. A maximum load of seven fingers of unripe plantain per cycle averaged 0.91 kg. The selection of unripe plantain for this study is in line with the WHO recommendation for diabetes patients, arising from current global health challenges and the exponential growth of chronic diseases [17].

$$\text{Machine Capacity (kg/h)} = \frac{W}{T} \tag{4}$$

where W is the weight of plantain fingers (kg), and T is the time taken to roast the plantain (h).

$$\text{Efficiency of the machine (\%)} = \frac{E_1}{E_2}\times 100 \tag{5}$$

where E_1 is the energy output, and E_2 is the energy input.

The force required to turn the weight of the plantain during roasting.

$$\mathrm{F} = \mathrm{m_1 g} \tag{6}$$

$$\mathrm{P^1} = \mathrm{m_2 g L} \tag{7}$$

where P^1 is the power requirement to turn the empty basket, m_2 is the mass of the rod, g is the center of gravity, and L is the length of the roasting basket.

$$= 100.33 \text{ W}.$$

$$P = m_2 gL + F \tag{8}$$

where P is the power requirement to turn the basket with plantain.

$$= 100.33 + (0.91 \times 9.8 \times 0.273)$$
$$= 102.77 \text{ W}.$$

Thus, the power requirement to turn the handle is 102.77 W. This falls within the sustainable human potential power of 70–500 W.

1.1.2 Sensory evaluation of results

Consumer assessment of the overall acceptability of the roasted plantain was carried out following the method described [15]. Twenty male and twenty female staff of Bells University of Technology, Ota, Nigeria, were selected randomly from regular consumers of roasted plantain for the evaluation. Coded samples were placed in separate identical tight polythene bags and assessed using a questionnaire among the forty respondents for color, flavor, hardness, moisture release, and chewiness on a hedonic scale of 1–5 points where: 1 = poor and 5 = excellent.

1.1.3 Performance evaluation results

From the experimental trials, the plantain roaster was observed to take 20 minutes for electric and gas sources and 25 minutes for the charcoal heat source to reach the desired temperatures of 200°C and 190°C, respectively, for the seven fingers of plantain per roasting cycle per compartment. The chamber temperatures were observed to drop by 20, 15, and 15°C, respectively, for the electric, gas, and charcoal sources after loading

Figure 2.
Snapshot of plantain roaster.

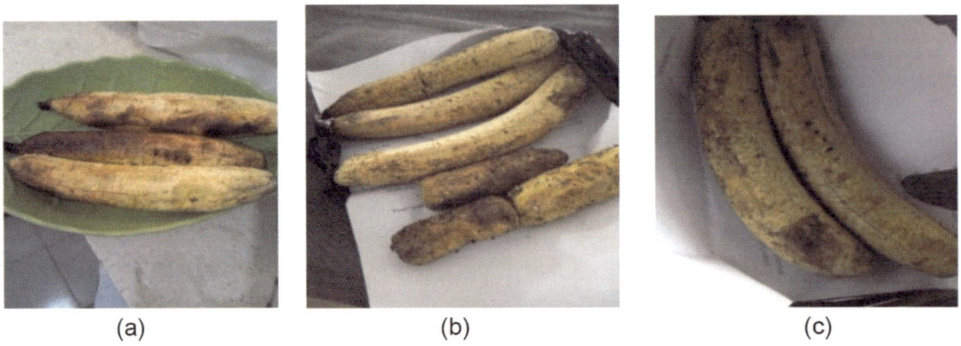

Figure 3.
Roasted plantain using a gas heat source (a), electric heat source (b), and charcoal heat source (c).

[14]. The plantain roaster is shown (**Figure 2**) below, with the color gradually changing to brown as shown in the figures a, b and, c for gas, electric, and charcoal heat sources, respectively. The interplay of heat by conduction on the roasted product (*booli*) as well as by convection and radiation has been well explained [14].

While the average capacity of the machine was 3.74 kg/h and the machine efficiency was 96.32% in each of the three compartments, radiation heat loss was 31.07, 31.07, and 29.23 W for the gas, electric, and charcoal heat sources, respectively; **Figure 3** below also shows the roasted products obtained.

The results of the performance evaluation of the roaster on the different heat sources are presented in **Tables 1–4** below.

From the heat transfer rates (**Table 1**) above, it is evident that over 60% of the heat supplied is derived from conduction. From the results of the proximate analysis (**Table 2**), moisture loss is the highest in the charcoal-roasted product, while sensory

Roaster compartment	Heat supplied (W)	Heat of conduction (W)	Heat of convection (W)	Heat of radiation (W)
Gas	846.28	538.02	277.19	31.07
Electric	846.28	538.02	277.19	31.07
Charcoal	793.60	506.37	258.00	29.23

Table 1.
Heat transfer rate for the roasting machine.

Test sample	Moisture (%)	Ash (%)	Crude fiber (%)	Carbohydrate (%)	Fat (%)	Protein (%)
Fresh	$58.27^a \pm 0.58$	$5.89^c \pm 0.13$	$2.83^a \pm 0.07$	$23.80^a \pm 0.24$	$2.16^a \pm 31$	$6.90^a \pm 16$
Gas	$46.26^b \pm 0.40$	$7.45^b \pm 0.53$	$3.27^b \pm 28$	$33.41^d \pm 0.38$	$4.52^c \pm 0.03$	$8.11^b \pm 0.15$
Electric	$45.94^b \pm 0.21$	$8.30^a \pm 0.12$	$3.41^b \pm 0.22$	$32.35^b \pm 0.45$	$4.68^b \pm 0.32$	$8.35^b \pm 0.14$
Charcoal	$39.59^c \pm 0.53$	$7.00^b \pm 0.05$	$3.43^b \pm 0.17$	$43.02^c \pm 0.14$	$4.64^c \pm 0.44$	$5.38^c \pm 0.30$

Data expressed as mean ± SD (n = 3). Means with different lowercase letters a, b, c, are significantly different (P < 0.05).

Table 2.
Proximate composition of fresh and roasted plantain.

Test sample	Potassium mg/100 g	Magnesium mg/100 g	Vitamin B6 mg/100 g
Fresh plantain	240.70[a]	111.35[a]	0.21[a]
Electric	309.43[b]	119.21[b]	0.26[b]
Gas	305.75[c]	119.87[b]	0.25[c]
Charcoal	312.10[d]	120.39[b]	0.27[d]

Means with different lowercase letters are significantly different ($P < 0.05$), $n = 3$.

Table 3.
Analysis of selected micronutrients in fresh and roasted plantains.

Oil	Groundnut	Corn	Soya
Weight of cup (g)	44.95	44.95	43.56
Weight of samples (g)	2	2	2
Weight after extraction (g)	45.44	45.36	43.99
Weight of oil (g)	0.49	0.41	0.43

Table 4.
Fat uptake of chin-chin after frying using electricity.

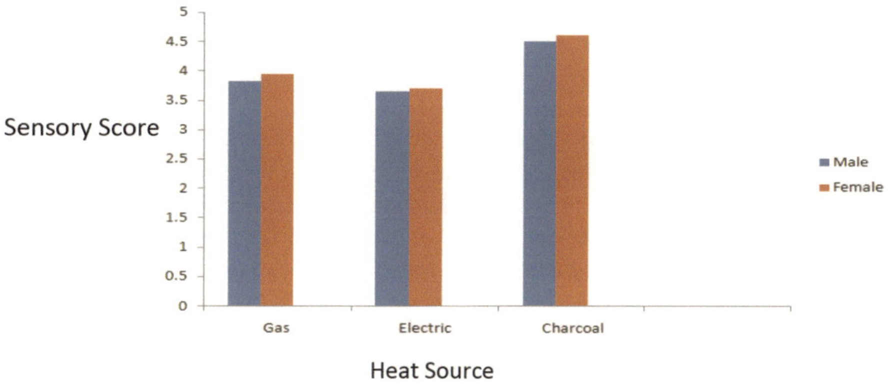

Figure 4.
Effect of roasting method on sensory attributes of plantain.

evaluation of the products (**Figure 4**) showed the preference of the panelists for the charcoal-roasted plantain, perhaps because gradual roasting influenced by prolonged residence time using the charcoal heat source produced the sweetest smelling flavor found the most acceptable in the product. The changes in selected micronutrients of the fresh and roasted products, namely, potassium, magnesium, and vitamin B6 are also presented (**Table 3**).

The sensory results showed a trend of higher scores from female panelists compared to males, suggesting a higher sensory acuity of the female panelists.

1.2 Frying

Deep-fat frying has been employed for centuries. Dogan et al. [18] noted that the Latin and Greek words for "frying" originated from the word "roasting," suggesting that frying may have developed from roasting. Today, it is an extremely popular technology employed in various kitchens all over the world and in about 85% of food service establishments for cooking chicken, fish, breaded vegetables, specialized pastries, French-fried potatoes, and other foods.

The simplest traditional process of deep-fat frying uses a kettle of oil heated on a stove or over an open fire in which small portions of the food are immersed in hot oil and removed when fried as determined by the experience of the cook. However, the first real breakthrough in the technology was the introduction of a continuous cooking process, which involves the immersion of the food in hot oil or fat for a specified period, followed by draining, cooling, and further processing or consumption [19].

The cooking medium during frying is the hot oil, also known as shortening, frying compound, or fat. The quality of the final food product largely depends on the quality of the oil. During the frying process, the inner moisture in the product is converted to steam, creating a pressure gradient. The surface dries out, causing the oil to adhere to the product surface and enter the surface of the damaged area [20]. Many factors have been reported to affect oil absorption in fried foods [21], including the quality and composition of the oil, temperature, time, product composition, moisture content, shape, porosity, pre-frying treatment, surface treatment, initial interface tension, and crust size [21].

The type of frying fat influences the quality of the finished product in terms of flavor, texture, shelf life, and nutritional attributes. Oils that have been exposed to a high temperature when left in the open air are subject to thermolytic and oxidative reactions [22]. These result in their partial conversion to volatile chain-scission products, nonvolatile oxidative derivatives, and dimeric, polymeric, or cyclic substances.

This is why the quality of fried foods depends not only on the type of food and frying condition but also on the nature of the oil used for frying. Selection of a stable frying oil is therefore of great importance in maintaining minimum product deterioration during frying and consequently a high-quality of the fried product [23].

Heating of the oil aids heat transfer by conduction and convection, the latter being caused by free water boiling at the surface upon immersion of the moist food in the hot oil. The moisture vaporizes and creates a path known as capillary pore, through which hot oil enters the food. The influence of oil uptake in this reaction is significant, with crust formation, shrinkage, and swelling occurring, thus inducing macro- and microstructural changes. This influences the vapor and liquid diffusion and safety assurance, and yields final products with the taste and textural characteristics expected by the consumer.

Typically, deep-fat frying is conducted at temperatures ranging from 120 to 180°C [24]. It is a complex process that involves simultaneous heat and mass transfer. It induces a variety of physicochemical changes in both the food and the frying medium. The principles underlying the mechanisms of oil absorption and water evaporation have been shown to be intimately related [25].

In fact, an extensive review of recent developments in the use of innovative methods for efficient frying of foods has been reported [26].

Various prototypes of deep fat fryers have been developed, including open fryers. These have been either single heat or double heat sources. Pressure fryers have also been designed to retain the vapor inside the fryer while cooling. The frying vessel captures the steam from the cooked food and increases the internal pressure until no more moisture is released from the food. The pressure usually ranges from 34473.80 to 82737.12 N/m^2 [20]. At such high pressures, it is not surprising that deep-fat fried products retain more flavor, and this constitutes a significant reason why they are widely popular across the globe. However, there are many areas where the energy source available may only be gas or charcoal; hence, the concept of a triple heat source has been developed in this study [15, 25].

1.2.1 Methodology

The desire for a deep-fat frying process equipment that is more user-friendly, efficient, and versatile led to the development of a simple, low-cost, and affordable multi-heat source deep fat fryer. Ogunmoyela et al. [8] reported the design and development of the multi-heat source deep fat fryer and its performance characteristics, as well as the methodology for determining oil uptake and vitamin A retention in the products, which are summarized below.

1.2.1.1 Design development and component characteristics of the multi-heat source deep fryer

The design, fabrication, and performance evaluation, including the major components as well as the controls of the fryer, have been described in detail in the literature studies [15, 25] and are as shown in **Figure 5**. The plain and front views, as well as the pictorial representations of the deep fat fryer, are also shown in **Figures 6–8**, respectively.

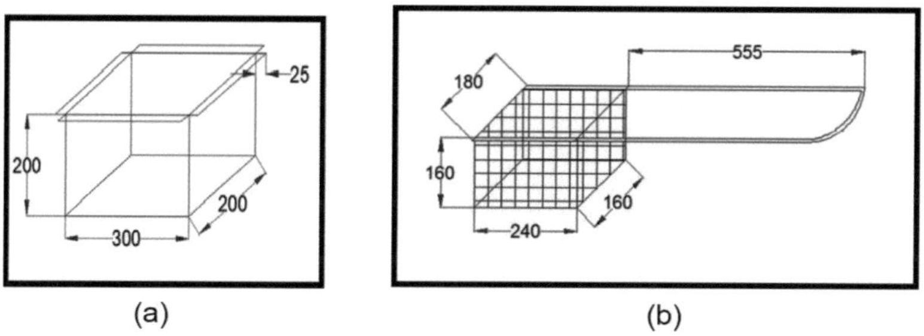

Figure 5.
Major components of the machine. A: Stainless steel frying pot, and B: stainless steel frying basket.

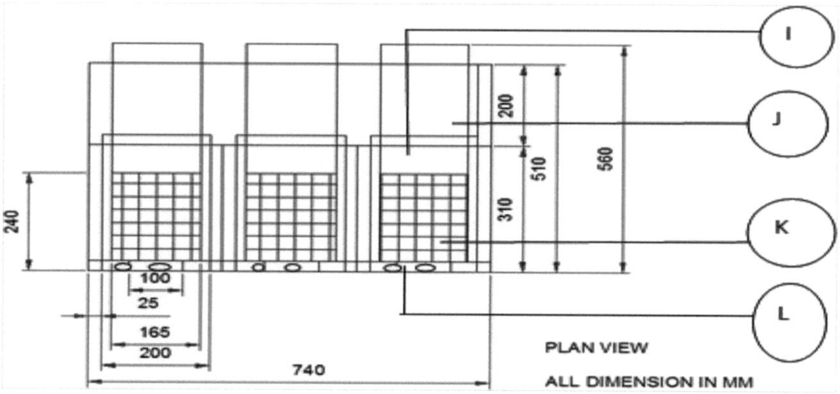

Figure 6.
Plain view of the machine. I: Stainless steel pot, J: basket lifter, K: stainless steel basket, and L: hinges.

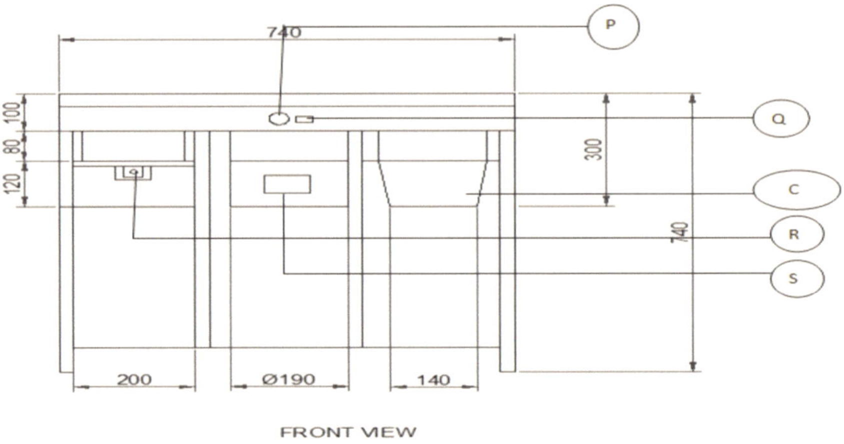

Figure 7.
Front view of the machine. P: Thermostat, Q: fan switch, C: charcoal compartment, R: gas control, and S: electric hot plate.

Figure 8.
Pictorial representation of the multi-heat source deep fat fryer (MSDFF).

The frame is divided into three rectangular but equal sections to facilitate a standing support to the pot hinged to the frame and connected to the frying pot for easy swing movements. The rectangular-shaped frying pot is made of stainless steel and sits on the frame. It can easily be removed for cleaning, while the frying basket is of the same shape, also made of stainless steel but smaller than the frying pot for free movement when hinged, apart from the need to allow free movement of hot air by convection.

The electric plate has a power of 1500 watts and is fitted with a sensor to detect changes in temperature, while the gas burner is welded to the frame support for the electric radiation plate. When the set temperature is attained, it cuts off automatically and vice versa. The charcoal compartment is also designed in this case to have a load-bearing capacity of 2 kg of charcoal for heating the fryer.

The fryer was developed using low-cost, locally available materials for affordability, at a total overall cost of about N50,000 (ca.$150.00) only but ensuring stainless steel for all contact surfaces to enable easy cleaning and to prevent corrosion and contamination.

The machine design considered the need for easy assembling, dismantling, adjustment, and operation, as well as the safety of the operator in the positioning of the various component parts. With a power requirement of 1.43 Kw, it is designed to be minimally efficient for processing any type of food material using good quality frying oils.

1.2.1.2 Performance Characteristics and Validation

The multi-heat source deep fat fryer was designed to use multiple heat sources, namely, charcoal, gas, and electricity, as sources of energy tested in different frying oils, namely, groundnut, soya, and corn oils. Heat is driven by conduction during deep-fat frying in the medium in line with the second law of thermodynamics. It is transmitted into the oil by convection and by conduction into the interior of the food. Part of it also escapes to the atmosphere by radiation since the process usually occurs in an open system.

The performance of the fryer was tested using wheat flour, sugar, eggs, baking powder, salt, and margarine mixed into a dough and fried with the different oils to obtain different products. Performance in each of these oils was tested using the different heat energy sources, namely, charcoal, gas, and electricity, respectively.

The fryer was successfully tested and validated in deep-fat frying of fried wheat flour chips, known as *chin-chin*; wheat flour balls known as *puff-puff*; as well as plantain chips, bean flour, and yam chips.

1.2.2 Performance Measurements and Process Chemistry

Deep-fat frying is a complex process that involves simultaneous heat and mass transfer. The process induces a variety of physicochemical changes in both the dough and the frying medium. As foods are hygroscopic and carry significant quantities of bound water in their porous matrices, the water diffuses from the matrix during frying, creating pathways usually referred to as "capillary pores." The formation of these capillary pores enhances oil absorption. As the food is fried, the moisture is converted to steam and released under pressure.

The type of product, process, and intensity of heating, coupled with the initial moisture content of the food product influences the final pore structure. Various studies have reported that the relationship between moisture loss and fat absorption is proportional and linear [27]. The diffusion rate of the moisture into the fat and that of the fat through the capillary pores depend on the temperature gradient across the heating medium. Since the two are proportional to one another, the basic equation for mass flux is applicable.

The mechanisms of water evaporation and oil absorption have been well described in the literature. As the product is immersed into the hot oil, the initial fat absorption takes place through surface wetting, by capillary action. As the product heats up, moisture is converted to steam, migrating to the surface and eventually into the frying medium due to a pressure differential. The vapor being released from the dough surface impedes the intrusion of fat into the product during surface boiling. Thus, the color of the dough gradually changes to brown (**Figure 9**) with heat conduction by the frying pot, transfer by convection within the hot oil, conduction into the interior of the food, as well as radiation heat losses.

The capacity of the machine using groundnut, corn, and soya oils (**Figure 10**) was found to be 6.90, 8.50, and 7.68 kg/h using electricity as the energy source; 8.50, 8.60, and 8.46 kg/h using gas as an energy source; and 5.60, 6.80, and 5.60 kg/h with charcoal as the energy source. From these results, the conductive heat requirement of the machine was found to be 1428 W, while the heat required for effective frying was 1392 W, and heat loss by radiation was only 36 W.

The results of the fat uptake (**Tables 4–6**) of the *chin-chin* fried with the fryer, using different compartments of electric, gas, and charcoal energy sources, respectively, are also presented.

The *chin-chin* fried with corn oil gave the least fat uptake followed by the soya oil when using gas. Corn oil also gave the highest vitamin A retention (**Figure 11**) given

F G

Figure 9.
F: Gradual changes during frying operation; G: chin-chin produced ready for packaging.

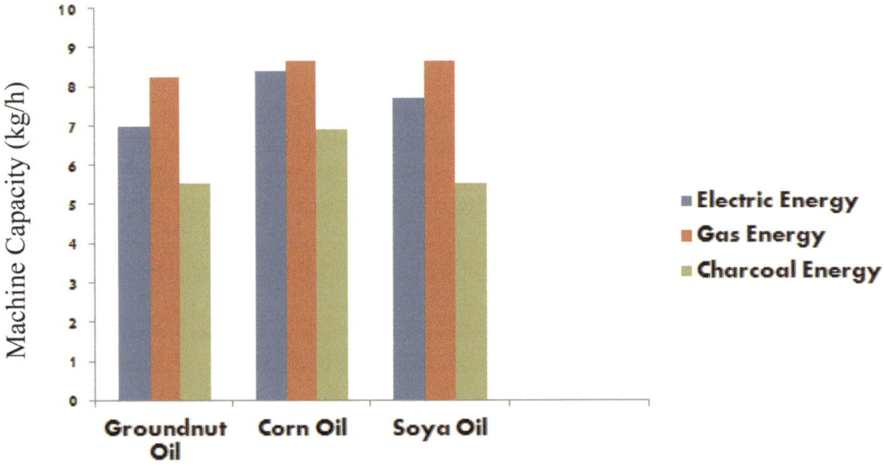

Figure 10.
Machine capacity (kg/h) using different oils and different heat sources.

Oil	Groundnut	Corn	Soya
Weight of cup (g)	44.43	44.71	44.83
Weight of samples (g)	2	2	2
Weight after extraction (g)	44.94	45.17	45.30
Weight of oil (g)	0.51	0.46	0.47

Table 5.
Fat uptake of chin-chin after frying using gas.

Oil	Groundnut	Corn	Soya
Weight of cup (g)	44.82	44.96	44.83
Weight of samples (g)	2	2	2
Weight after extraction (g)	45.49	45.53	45.46
Weight of oil (g)	0.67	0.57	0.63

Table 6.
Fat uptake of chin-chin after frying using charcoal.

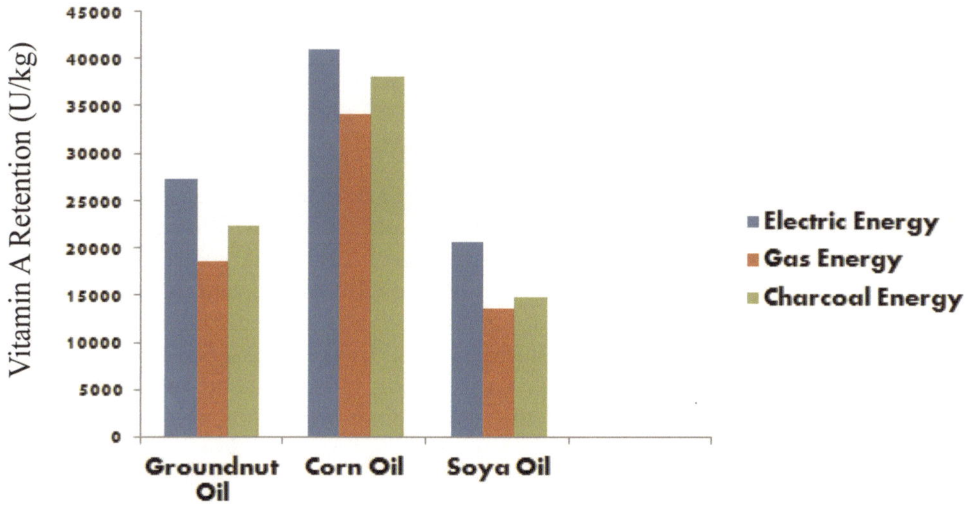

Figure 11.
Vitamin A retention in chin-chin during frying using different heat sources.

the mandatory vitamin A level of 3000 IU/kg in flour and 2000 IU/kg in vegetable oils under the National Food Fortification Programme in Nigeria.

Given the implications of high fat intake on the increasing incidences of cardiovascular diseases, the figures obtained here indicate the significant promise of the use of this equipment in the control of fat uptake in deep-fat fried products.

1.3 Smoking

Smoking is the practice or process of seasoning, to preserve food either by exposing it to smoke from a burning or smoldering substance, usually wood, or by cold smoking at a reduced temperature of 12–25°C. This is to impart flavor and ensure the adequacy of preservatives added to the product. Smoking can also be done by bringing food into contact with vaporized liquid smoke. In many African countries, smoking is the most significant method for preserving fish and wildlife. Wood smoke is made up of a variety of organic chemical components, some of which have antibacterial properties [28]. When wood smoke is condensed into water, it produces liquid smoke, which can be utilized for food smoke flavoring. The dangers associated with the illegal use of chemically preserved wood are one of the concerns associated with smoked foods, and this is why gas systems have been advocated in recent years by developed countries. In fact, the current trends of using green technologies in food production and processing have been reported [29]. In particular, a recent review has shown the technological advances in Ghana which led to the development of such smoking methods as the FAO-Thiaroye technique of processing (FTT), and Abuesi gas fish smoker for fish smoking and drying, with results of lower PAH4 levels and uniform appearance of end products [30]. However, there is no doubt that wood-smoked foods are still preferred in many developing countries of the world, and they are safe if they are made from fresh raw materials that are free of natural toxins, chemical pollutants, pathogens, and parasites; and if the storage conditions do not promote microbial proliferation or toxin production [31]. In addition, the sensory characteristics of such products are usually more intense and better than those of gas-smoked products.

Thus, there is no doubt that smoking is a unit operation that has gained overwhelming acceptance across the globe. But the conventional methods for smoking are often laborious, stressful, and unhygienic, thus posing health risks to processors. Based on these limitations, more modern, low-cost, and effective smoking kilns suitable for small- and medium-scale quality production of fish and meat have been investigated in various studies.

Smoking is a combination of salting, drying, and heating of fishery products. The regulation of physicochemical parameters such as pH, VBN, TBARS, fatty acid content, and texture profiling using the smoking preservation process increases sensory quality, allowing for longer storage times of high-quality fish products. Smoking of the food considerably slows down oxidative changes and prevents microbial growth. When smoke from incomplete combustion of wood or sawdust is deposited on the surface of processed fish, volatile chemical compounds are released, which help to suppress bacterial growth [32]. Due to the unique color and flavor, smoked items have a high demand among consumers. It has been reported that smoking of wood or sawdust releases a variety of complex chemicals such as phenols, ethers, esters, hydrocarbons, acids, alcohols, and ketones, which are responsible for the subsequent color and flavor development [33].

Smoking has also been classified into three types based on temperature: cold (12–25°C), warm (25–45°C), and hot (40–100°C). The type of heating procedure selected is critical to product quality. However, heating can promote protein denaturation in such products, resulting in reduction of both nutritional and functional qualities. To obtain a premium-grade smoked product, optimizing the time, temperature, and sawdust material in the hot smoking procedure has therefore been shown to be vital [34]. Sensory evaluation (color, texture, odor, flavor, and overall acceptance), physicochemical assessment (pH level, VBN level, TBARS level, and TMAO and fatty acid content), and microbial growth are also important parameters for establishing the qualities of smoked fish products [34]. Many smoked products have, however, been found to be mutagenic and carcinogenic due to polycyclic aromatic hydrocarbons (PAHs) found in wood smoke. In recent years, wood-smoked foods have been increasingly investigated for potential genotoxicity and carcinogenicity. However, it has been reported that PAHs are processed by enzymes in the human body, resulting in premutagenic and carcinogenic DNA adducts [35, 36]. Nevertheless, it is important to stress that these concerns are associated with improperly wood-smoked products, which is why the smoking process requires standardization, rather than being discarded as has been advocated in some developed countries.

In many regions of the world, fish processing *via* hot smoking or kiln has been practiced for centuries. In fact, smoked fish is one of the most popular delicacies in many developing countries, with various types of smoking kilns ranging from traditional open-fire to mud-brick and cylindrical drums.

Despite this fact, the absence of suitably controlled process has long been a challenge to the quality of smoked fish from many developing countries. Locally accessible technologies like mud bricks, stone, and firewood are commonly employed, but with negative impact on the quality of the finished product. The market value of the product declines due to smoke damage and unappealing appearance of processed fish, while standard conditions for quality assurance and hygiene remain a challenge [37].

To reduce the stress, drudgery, and health risks to processors associated with the conventional methods of smoking, a low-cost and effective smoking kiln suitable for small- and medium-scale quality production of fish and meat was developed and evaluated.

1.3.1 Design characteristics of low-cost smoking kiln

The smoking kiln is a double-jacketed cabinet with a thickness of 15 mm. It is made from mild steel and lagged with a fiber glass insulator purposely to prevent heat loss to the environment during smoke drying. The smoking chamber consists of set trays arranged into three rows and a smoking rack with the same length and breadth. The overall dimension of the cabinet is $600 \times 515 \times 650$ mm, and the dimension of the trays is $425 \times 325 \times 150$ mm. The fabrication and assembly of the smoking kiln was done at our Mechanical Engineering workshop of Bells University of Technology, Ota, Nigeria.

The trays have a trough fabricated to their ends to allow for a flow out of food product drippings during smoking without accumulation. This trough is connected to a pipe that runs from the top to the bottom at the back of the smoking kiln through which the troughs of the other trays connect, collecting all product drippings and expelling them to the outer part of the smoking kiln. The heat source of the smoking kiln is charcoal, which is contained in two pots, each with a dimension of $484 \times 120 \times 70$ mm, placed by the sides of the rack system that carries the trays.

The design allows for air circulation by convection in the combustion chamber with heated air carried in all directions of the loaded trays, and air inlets at the lower front of the smoking kiln facilitate the flow of heat. The chimney is fitted with an adjustable valve that controls the heat buildup within the smoking kiln and conducts the smoke to the external environment. The isometric and orthographic projections of the smoking kiln assembly (**Figures 12** and **13**) are presented below.

The machine was designed with a tray arrangement at the center of the kiln, to allow for proper air circulation *via* an indirect mode of heating. The region between the charcoal pots was perforated to allow the inflow of fresh air to support the combustion and mobility of smoke in the kiln. A tray system slightly sloped backward was adopted to allow oil leaching from product flow into a trough where it is collected.

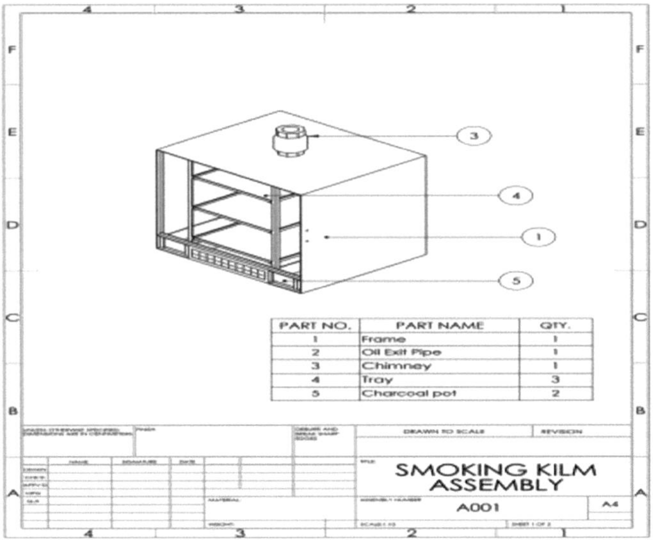

Figure 12.
Isometric projection of smoking kiln assembly.

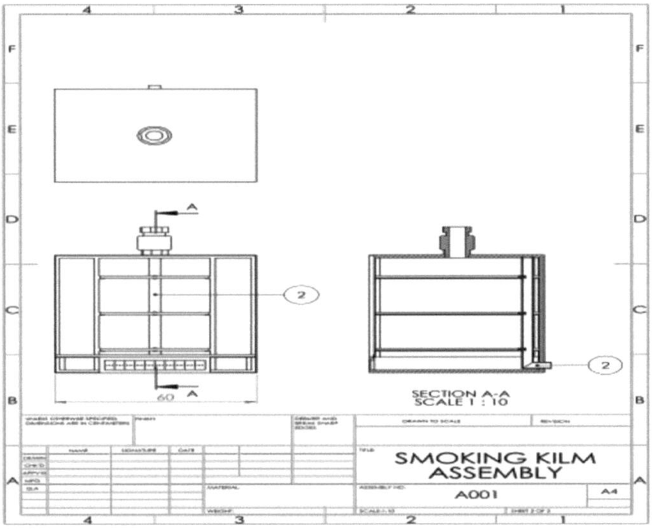

Figure 13.
Orthographic projection of smoking kiln assembly.

1.3.2 Design calculations

Volume of the fish tray (VT) is calculated as:

$$VT = l \times b \times h \tag{9}$$

where l = length of the tray = 42.5 cm, b = width of the tray = 32.5 cm, and h = height of the tray = 150 cm

Volume of the charcoal pot (VC): is calculated as:

$$VC = l \times b \times h \tag{10}$$

where l = length of charcoal pot = 48.4 cm, b = width of charcoal pot = 12 cm, and h = height of charcoal pot = 7 cm.

Determination of heat transfer by conduction, convection, radiation, and thermal resistance is in accordance with Ref. [38].

Heat transfer (q) by conduction is obtained as:

$$q_{cond} = \frac{-KA(T_2 - T_1)}{l} \tag{11}$$

where K = thermal conductivity of the material (45 W/m °C), A = area of the fish tray (0.41 m^2), l = thickness of fish tray (0.0015 m), T_1 = temperature of the inside smoking kiln (280°C), T_2 = temperature of the inside smoking product (90°C), while negative sign of K is to take care of the decrease in temperature along the direction of heat flow.

Heat transfer (q) by convection is obtained as:

$$q_{conv} = hA(t_s - t_f) \tag{12}$$

where h = coefficient of convective heat transfer (free convection) (20 $W/m^{2o}C$), A = area of the charcoal pot (0.12m^2), T_s = surface temperature (280°C), and T_f = fluid temperature (90°C).

Heat transfer (q) by radiation is obtained as:

$$q_{rad} = F\delta A\left({T_1}^4 - {T_2}^4\right) \tag{13}$$

where F = emissivity coefficient of mild steel (0.20), δ = Stefan Boltzmann's constant (5.67×10^{-8}), A = area of the charcoal pot (0.12 m^2), T_1 = temperature of the inside smoking kiln (553 K), and T_2 = temperature of the outside smoking kiln (303 K).

Heat capacity q_{total} of the machine is obtained by:

$$q_{total} = q_{cond} + q_{conv} + q_{rad} \tag{14}$$

Value of thermal resistance is obtained by:

$$(R_{th})_{rad} = \frac{T_1 - T_2}{q_{rad}} \tag{15}$$

Heat required for smoking products:

$$q = M \times C_p \times \Delta T \tag{16}$$

where M = mass of sample in the smoking kiln at a time (fish = 5.20 kg, beef = 5.85 kg, and chicken = 8.20 kg), ΔT = change in temperature (°C), and C_p = specific heat capacity of products: fish = 3.60 kJ/kg°C, beef = 2.85 kJ/kg°C, and chicken = 3.22 kJ/kg°C [39].

The heat capacity q_{total} of the machine is given as 2337.57 kJ, and this exceeds the heat requirement for each of the products smoked: fish = 1123.20 kJ, beef 1000.35 kJ, and chicken = 1584.24 kJ.

1.3.3 Performance Evaluation of results

Heat capacity of the machine was calculated to be 2337.57 Kw, while the heat requirements for smoking fish, beef, and chicken were calculated to be 1123.20, 1000.35, and 1584.24 kJ, respectively. The rates of moisture removal (**Table 7**) in the smoking kiln were 25.87%/hr., 23.37%/hr., 28.05%/hr., and 24.51%/hr. for Atlantic mackerel, herring fish, beef, and chicken, respectively. Smoking temperature was determined to be 90°C, which was in accordance with the findings of Rahman [40] that the smoking temperature range suitable for effective drying is 80–90°C. It was observed that at the various tray levels of the smoking kiln, there was a slight temperature difference, probably due to hot air being of lighter density than cold air and floating upward. In the smoking kiln, since the heat source is not directly under the rack system, it floats to the upper part of the smoking kiln, and the products at upper tray dry faster. The performance results for different products are summarized in **Tables 7** and **8** below.

The results of the sensory evaluation of different products in the machine (**Table 8**) also showed that there was no significant difference ($p > 0.05$) except for the texture. Although there was no significant difference in overall acceptability of the samples, there was a clear preference for Atlantic mackerel fish. However, at 5% level of significance, panelists did not find any significant difference ($p > 0.05$) in the sensory attributes of the smoked products. **Figure 14** shows an exploded view of the kiln.

Samples	Time taken (hrs)	Initial weight (kg)	Final weight (kg)	Moisture loss (%)	Rate of moisture removal (%/hr)
Beef	2	5.85	2.57	56.10	28.05
Atlantic mackerel	2	5.20	2.51	51.73	25.87
Chicken	2	8.20	4.18	49.02	24.51
Herring fish	2	5.20	2.77	46.73	23.37

Table 7.
Machine evaluation.

Sample	Appearance	Aroma	Taste	Texture	Overall acceptability
SMBE	6.95 ± 1.36^a	7.35 ± 1.28^a	7.45 ± 0.83^a	7.10 ± 1.41^a	7.65 ± 0.88^a
SFTI	7.50 ± 1.10^a	7.75 ± 1.16^a	7.95 ± 1.00^a	8.10 ± 0.79^{ab}	8.00 ± 0.79^a
SMCH	7.07 ± 1.08^a	7.30 ± 1.30^a	7.50 ± 1.28^a	7.40 ± 1.47^{ab}	7.65 ± 1.04^a
SFSH	7.45 ± 1.28^a	7.50 ± 1.19^a	7.55 ± 1.19^a	7.45 ± 1.19^b	7.50 ± 1.15^a

SMBE: smoked beef, SFTI: smoked Atlantic mackerel fish, SMCH: smoked chicken, SFSH: smoked herring fish. Values are means ± standard deviation of duplicate determinations. The mean values of the samples within a column with different superscripts (letters) are significantly different ($p < 0.05$).

Table 8.
Sensory evaluation of smoked samples.

Figure 14.
Exploded view of the smoking kiln.

1.4 Fermentation

West African food cultures, like many other parts of the world, are rich in spontaneously fermented foods, the majority of which have been passed down from one generation to another. The fermentation process involves the conversion of starch or sugar into an alcohol or acid. This is the basis of several foods, including baked products. During the fermentation of wheat flour dough, for example, carbon dioxide is produced and trapped as pockets of air within the dough, causing the dough to rise. During subsequent baking, the carbon dioxide expands and causes the dough to rise further, with the alcohol produced evaporating during this baking step.

The fermentation of foods in West Africa is said to account for 40% of the population's diet, a percentage that increases with decreasing income. Africans usually ferment cereal-based foods such as sorghum, millet and maize; roots such as cassava; fruits; vegetables; and less commonly meat and fish [41]. Fermentation also covers leguminous plants and oilseeds, to produce fermented condiments that are used as flavorants in soups. Such fermented condiments include "*ogiri*" from castor bean (*Ricinus communis*), "*dawadawa*" from African locust beans (*Parkia biglobosa*), and "*ugba*" from African oil bean (*Pentaclethra macrophylla*).

One of the major advantages of fermentation is the enhancement of both nutritional and sensory quality of foods by the conversion of macronutrients such as proteins and sugars into easily digestible compounds and the development of flavour compounds. Fermented African locust bean seeds, for example, are a rich source of protein and consist of oil, dietary fiber, vitamins (vitamin B, riboflavin, and vitamin A) and minerals. The most common groups of microorganisms involved in food fermentation are bacteria, yeasts, and mold [42]. Spices and condiments are plant-derived substances (from dried seeds, fruit, root, bark, and leaves) used in minute quantities as food additives to stimulate flavor and taste in foods, beverages, and drugs; improve color; and in some cases, serve as preservatives [43] and overall sensory acceptability of foods. Condiments are applied in the form of a sauce powder to contribute calorie and protein intake and are generously added to soups as low-cost meat substitutes by low-income families [44]. Thus, the awareness of the benefits of eating food products with little or no chemical food additives or preservatives for healthy living and life expectancy is increasingly being promoted all over the world. This is why microbial fermentation technology has become such a promising, rapidly growing, revolutionary field involving the use of microbes for the production of compounds that are of immense use in the production of biofuels, pharmaceutical, environmentally friendly materials, energy, fine chemicals, and the food industry in the quest for a bio-based society, with an eye on the sustainability factor [45].

Foods prepared with chemical additives and preservatives are susceptible to chronic and noncommunicable diseases. Therefore, this global shift to naturally processed foods calls for indigenous food processing techniques that will guarantee consumer safety, healthy living, and storability. An example of recent developments in the packaging and commercialization of fermented African locust beans to cube form using a locally fabricated machine is presented in this chapter. The production and cubing of African locust beans using a prototype cubing machine fabricated in the Mechanical Engineering Workshop of Bells University of Technology, Ota, in Nigeria is described below.

Using the traditional fermentation process, the dried locust beans are inspected, and removal of immature seed and broken and damaged locust beans takes place to avoid poor quality and unsafe finished products. Prewashing is done before the locust beans are placed in clean pots. They are then allowed to boil for 6 hours so as to partially de-shell the locust beans. The boiled locust beans are then drained and allowed to cool for easier mashing. The cooled locust beans are placed in the mortar and hand-mashed with the addition of potash. The de-shelled locust beans are thereafter hand-sorted, washed with water and placed in a calabash containing washed banana leaves, and then allowed to ferment for 24 hours. In order to facilitate the fermentation process, a little amount of salt is added. The locust beans are then dried using hot air oven at 60°C for 24 hours to reduce the bulkiness by 29%. The dried locust beans ware dry-milled with using a hammer mill, and the resulting milled locust beans are sieved to obtain a homogenous size.

The locust beans' powder is then divided into two equal parts. One of them is subjected to cubing using the prototype cubing machine with addition of lecithin to serve as the binding agent in ratio 3:1 (locust bean powder: lecithin). The powdered and cubed locust beans are then packaged for further analyses. In the design of prototype cubing machine, various factors were considered in the selection of material, including material availability, suitability, durability, and cost of materials to meet the desired quality performance. The process flow diagram for the production of fermented, dried, and cubed African locus beans is shown (**Figure 15**) below.

1.4.1 Prototype cubing machine design

The major components of the prototype machine (**Figure 15**) are u-beam, cubing unit, mold, cylinder, and the frame. The u-beam made from mild steel was cut and shaped into 1500 mm × 210 mm × 26 mm; the cubing unit consists of cutting blade made from stainless steel and a cylinder spring and handle; the mold houses the blade and is made from stainless steel with dimension 110 mm × 100 mm × 21.5 mm and capable of producing 200 cubes per hour; the cylinder consist of iron rod that is

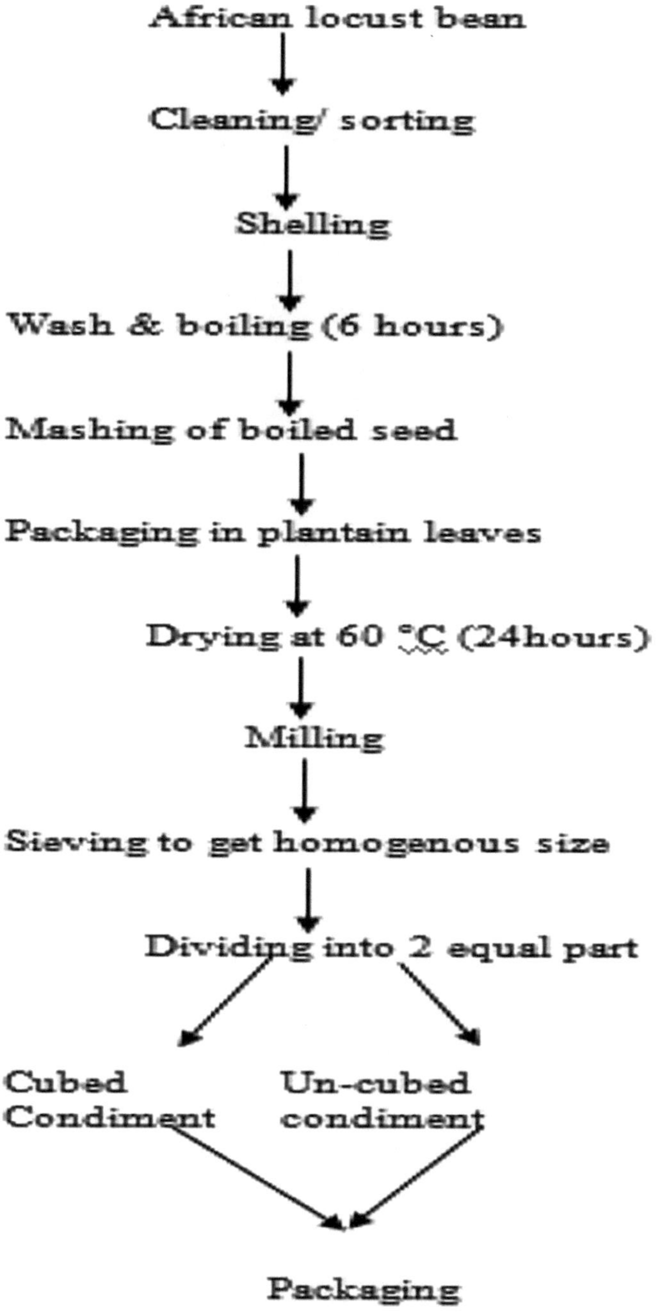

Figure 15.
Flow diagram for the production of fermented, dried, and cubed African locust beans.

10 mm tall and 2 mm in radius and made from mild steel. The outer part of the cylinder is lined with flat metal connected to the handle, and as the handle moves downward, it presses the spring, while the mold is in contact with the condiment on the table to form cubes. The frame unit made from angle iron provides support to other components of the machine. The upper part of the frame is the working table

Figure 16.
Snapshot of the prototype cubing machine.

that has direct contact with food made from stainless steel. The snapshot of the machine is shown below in **Figure 16**.

1.4.2 Design calculations and performance evaluation

The design analysis for the spring support (**Figure 17**), power requirement (**Figure 18**), machine capacity, and efficiency are considered as follows:

Selected design data are as stated:

Spring free length, L = 120 mm, Diameter of the wire, d = 2.5 mm, Number of active coils, N = 14, Diameter of the coil, D = 28 mm, Mass of the handle, m = 170 g, Mass of the connecting rod = 550 g, Allowable stress (δ_s) induced in the spring due to twisting for industrial spring = 6.25×10^5 Nm^{-2} [46].

The compression load of the spring, F was obtained using the relationship reported [47].

$$F = \delta_s 2\pi r L \tag{17}$$

where r = $\frac{d}{2}$.

$$F = 6.25 \text{ x } 10^5 \text{ x } 2 \text{ x } \frac{22}{7} \text{ x } 0.00125 \text{ x } 0.12.$$

F = 589 N.

Selected weight of the handle,

$$W = mg. \tag{18}$$

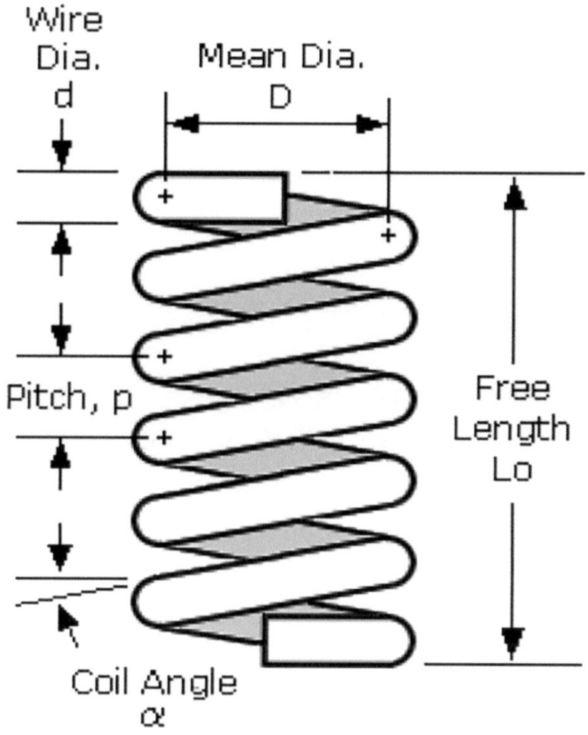

Figure 17.
Design for spring support.

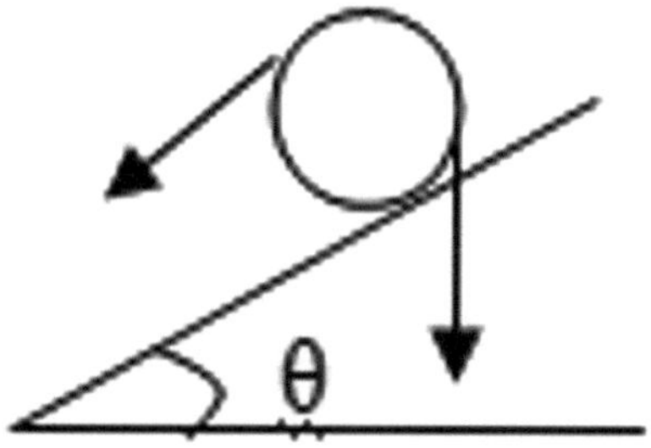

Figure 18.
Design for machine power requirement.

where m is the mass in kg, and g is the acceleration to gravity.

Therefore, W = 0.170 × 9.81.

W = 1.67 N.

Machine compressive force,

$$F_s = F + W \tag{19}$$

F_s = 589 + 1.67.

F_s = 590.67 N.

The maximum deflection of the spring, E was obtained from the relationship given in Ref. [48].

$$L - E = (n + 2)d \tag{20}$$

where n is the number of active coils.
E = 0.12 – E = (14 + 2) 0.0025.
E = 80 mm.
Therefore, the maximum deflection of the spring is 80 mm.
The torque T required to overcome friction in spring was obtained using the relationship as stated in Ref. [47].

$$T = \frac{\pi \delta_s d^2}{16K} \tag{21}$$

where K is stress concentration factor = 1.225.

$$T = \frac{\frac{22}{7} \times 6.25 \times 10^5 \times (0.0025)^2}{16 \times 1.225}$$

T = 0.63 Nm.
Therefore, the torque require to overcome friction in spring is 0.63 Nm.
Stiffness of spring was obtained from the relationship given in Ref. [46].

$$\sigma = \frac{\pi R^4 G}{2L} \tag{22}$$

where σ = stiffness of spring in Nm^{-1}.
R = radius of coil in m.
G = mild steel modulus of rigidity = 8.2 x 10^6 Nm^{-2} [49].

$$\sigma = \frac{\frac{22}{7} \times (0.014)^4 \times 8.2 \times 10^6}{2 \times 0.12}$$

σ = 4.13 Nm^{-1}.
Therefore, stiffness of the spring is 4.13 Nm^{-1}.

$$\mu = \text{Tan}\theta \tag{23}$$

where μ is the coefficient of friction, and θ is the angle of deflection

$$\text{Tan}\theta = \frac{x}{y} \tag{24}$$

where x is the maximum compressive force, and y is the weight of handle + weight of connecting rod.

$$\text{Tan}\theta = \frac{590.67}{1.67 + 5.40}$$

$$\text{Tan}\theta = 83.55$$

$$\theta = \text{Tan}^{-1} 83.55$$

$$\theta = 89^o$$

To calculate angular speed, S

$$S = \frac{r\theta}{t} \tag{25}$$

where θ is in radian, r is the length of plate base = 100 mm, and t is the time of deflection = 1 second.

$$S = \frac{0.1 \times 89\pi}{180}$$

S = 0.16 ms^{-1}.

To calculate machine power requirement for cubing, P

$$P = F_s S \tag{26}$$

P = 590.67 × 0.16.

P = 94.51 W.

Conclusively, the power required by the machine for effective cubing falls within sustainable human potential power is 70–500 W.

Design for machine efficiency

$$Efficiency = \frac{Output}{Input} \times 100 \tag{27}$$

During machine evaluation, five batches in triplicate were carried out, and average value for each batch was recorded (**Table 9**). The weight of the cubes was determined by analytical weighing balance, and time taken for each experiment was taken using a stopwatch.

Machine capacity is calculated as follows:

$$Capacity = \frac{N}{t} \tag{28}$$

where N is the number of cube produced, and t is the time taken in the production.

Average machine capacity = 191 cubes/hr.

Input (g)	Output (g)	Eff. (%)	Number of cube produced	Time taken (sec.)
500	422	84.40	84	1500
400	358	89.50	72	1320
350	326	93.14	65	1200
300	290	96.67	58	1110
250	246	98.40	49	1020

Average machine efficiency = 92.42%.

Table 9.
Machine performance evaluation.

1.4.3 Product weight control

The weights of fermented African locust bean cubes produced were found to be slightly different from one another even from the same batch production. This variation could be from the mixing or machine settings among others, and further improvements in machine performance are possible with the investigation of the actual cause of the problem using appropriate statistical control chart design.

Procedure for $\overline{\overline{x}}$ control chart

Below is the step-by-step procedure to determine the control limit using $\overline{\overline{x}}$ chart

i. Draw a sample $\{x_1, x_2, x_3, \ldots \ldots \ldots \ldots x_k\}$ of size k at a stage of the production process.

ii. Repeat (i) for n samples at equal intervals of time.

iii. Calculate the sum $\sum x$ for each sample.

iv. Calculate the mean $\overline{x} = \frac{\sum x}{k}$ for each sample.

v. Calculate the mean of the mean () in (iv) for all observations where $\overline{\overline{x}} = \frac{\sum -x}{n}$.

vi. Calculate the variance $s^2 = \frac{\sum (x-\overline{x})^2}{k}$ for each sample.

vii. Calculate the standard error (S) for every S^2 in (vi)

viii. Calculate the mean $-s = \frac{\sum s}{n}$ for all standard errors in (vii)

ix. Obtain the control limit for x as follow:

a. $\overline{\overline{x}}$ - Central control limit.

b. $\overline{\overline{x}} + \frac{3\overline{s}}{\sqrt{n-1}}$ - Upper control limit.

c. $\overline{\overline{x}} - \frac{3\overline{s}}{\sqrt{n-1}}$ - Lower control limit.

Table 2 shows triplicate weight readings of cubed fermented African locust beans for each batch production as illustrated for the determination of the control limits.

From **Table 10**,

$$\overline{\overline{x}} = 5.012$$

$$\overline{S_i} = 0.0162$$

Control limits:
Central control limit = 5.012.
Upper control limit = 5.024.
Lower control limit = 4.989.

Batches	Cube weight (g)						
	1	2	3	$\sum x_1$	$\overline{x}$	s^2	S
1	5.01	5.02	5.03	15.06	5.02	6.70×10^{-5}	0.0082
2	4.98	4.99	5.03	15.00	5.00	4.67×10^{-4}	0.0216
3	4.99	5.02	5.04	15.05	5.02	4.33×10^{-4}	0.0208
4	4.98	5.00	5.02	15.01	5.00	2.67×10^{-4}	0.0163
5	5.01	5.01	5.04	15.06	5.02	2.00×10^{-4}	0.0141
					25.06		0.081

Table 10.
Statistical control of batch production of fermented, dried, and cubed African locust beans.

1.4.4 Performance evaluation results

For the machine design analysis, some parameters were selected, while some were calculated using mathematical relationships. Those that were calculated are: compression load of the spring, F = 589 N; machine compressive force, Fs = 590.67 N; maximum deflection of the spring, E = 80 mm; torque, T = 0.63 Nm; stiffness of the spring, σ = 4.13 Nm^{-1}; angle of deflection of the machine, Θ = 89°; the angular speed of the machine, S = 0.16 ms^{-1}; machine power requirement, P = 94.51 W; average efficiency of the machine = 92.42%; and the average capacity of the machine = 191 cubes/hour. However, as the input of the machine increased from 250 to 500 g, the efficiency also decreased from 98.40 to 84.40%, while machine capacity increased from 175 cubes/hr. to 200 cubes/hr.

This means that the higher the input, the longer is the residence time and the rate at which efficiency drops, implying that more losses could be incurred. From the cube control chart (**Figure 19**), it is evident that in all the batches, the production process is

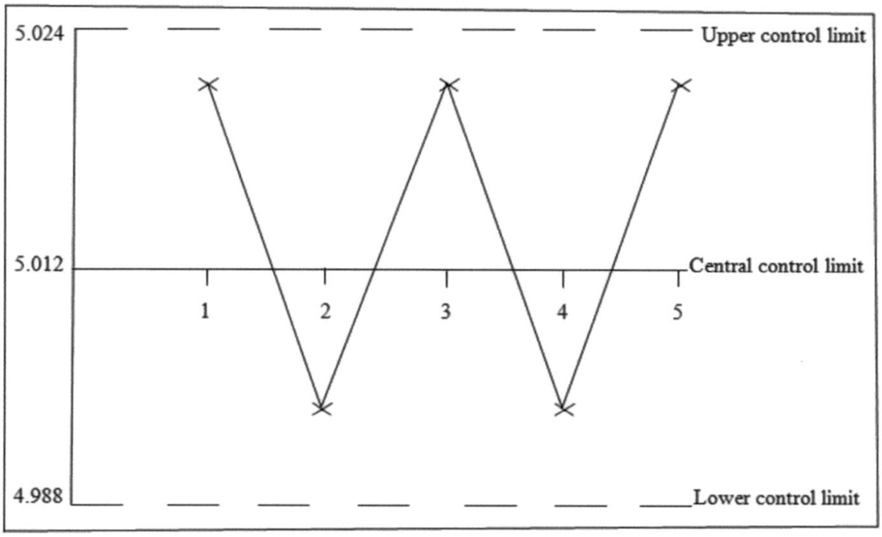

Figure 19.
Production control chart.

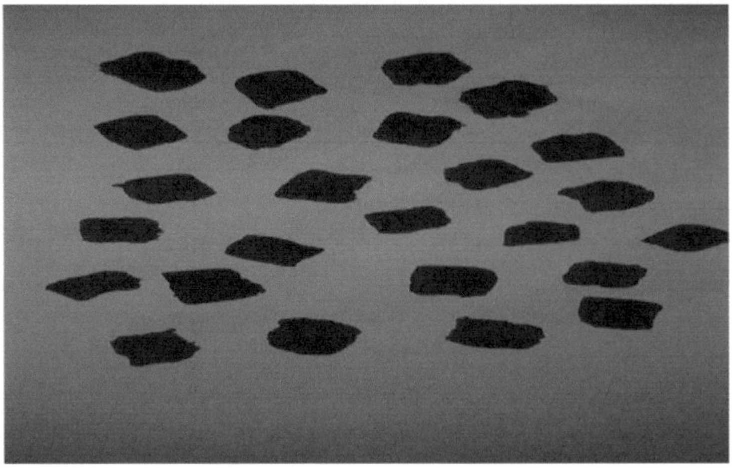

Figure 20.
Cube production from different batches.

in a state of statistical control. No significant assignable variation is detected. Slight differences observed in the weight of the product are due to inherent chance variations that are inevitable in any production process. Cubes produced from different batches as they are released from the cubing unit are shown (**Figure 20**).

1.5 Effects of processing on product quality

The results of proximate analysis of the cubed and un-cubed powder (**Table 11**) show a rapid decrease compared with 42.65% reported in Ref. [50]. The reduced moisture content could be the result of the drying process at 60°C for 24 hours, and this might improve the storability of the condiment. The crude fat content of the un-cubed sample was 12.13%, while that of cubed sample was 12.12%. These values are closely related to 10.65% reported in Ref. [51]. The results show that the values of crude protein for un-cubed African locust beans sample was 80.25% while that for cubed sample was 80.18%. The result of 37.32% obtained for the protein content of naturally fermented African locust beans is, however, in line with that reported in Ref. [50], while the increase in the fiber content of the cubed sample over the un-cubed sample could be traced to be an effect of the hydrophilic and lipophilic tendency of lecithin used as binding agent [52].

Nutrient	Un-cubed locust beans (%)	Cubed locust beans (%)
Moisture content	6.43	5.43
Crude fat	12.13	12.12
Crude protein	80.25	80.18
Crude fiber	0.48	0.86
Total Ash content	0.61	0.92
Carbohydrate content	0.12	0.50

Table 11.
Proximate composition of un-cubed and cubed fermented African locust beans.

1.6 Processing Losses

The process of transforming raw ingredients into food through any of these technologies usually leads to nutrient losses. These losses may arise due to sensitivity to heat, light, oxygen or pH of the solvent, or a combination of these. Food fortification, which is the addition of one or more essential nutrients to food, is often employed as one way of restoring such losses occurring during processing or correcting a demonstrated deficiency in the population. Vitamin A, iodine, and iron remain three of the most important micronutrients from a public health perspective, and wheat flour, sugar, vegetable oils, and salt are some of the commonest vehicles for carrying these nutrients in food products. We have studied in our laboratories, in particular, the effects of baking and frying on the vitamin A and iron contents of pan bread and doughnuts [53]. The results obtained showed slight variations in vitamin A and iron levels of treated and market flour samples (**Table 12**), vitamin A content of pan bread (**Table 13**), vitamin A content of fried doughnuts (**Table 14**), as well as iron levels of baked pan bread (**Table 15**) and fried doughnuts (**Table 16**), respectively.

From these results (**Tables 13–16**), the baking process at 175°C for 45 minutes using dough proofed for 60 minutes retained more vitamin A than deep-fat frying at 185^0 C for 5 minutes using dough proofed for 45 minutes, while process losses of iron were comparable [54]. Nevertheless, a 25–35% vitamin A loss was recorded during baking compared to a 33–40% vitamin A loss recorded after deep-fat frying.

Clearly, these results confirmed that iron is more stable under various processing conditions except in the presence of moisture. Similar baking and deep-fat frying studies also showed that during baking at 175°C for 45 minutes with dough proofed for 60 minutes, only a 6–10% loss of iron was recorded. The iron retention was better in dough proofed for 90 mins compared with deep-fat frying at 185°C for 5 mins where a 6–15% loss in doughnuts was obtained.

Flour type	Vitamin A content (2 g)	Iron levels (ppm)
Blank	6000 IU	58
Treated	35,000 IU	66
Market	27,500 IU	68

Table 12.
Vitamin A and iron levels of blank, treated, and market wheat flour samples.

Flour type/Proofing time (mins.)	Vitamin A level before baking (IU)	Vitamin A level after baking (IU)	% loss
Blank: 30	5000	3500	30
60	4000	3000	25
Treated: 30	27,500	18,000	35
60	33,000	24,000	27
Market: 30	24,000	16,000	33
60	28,000	22,000	21

Table 13.
Vitamin A content of pan bread before and after baking at 1750C for 45 minutes.

Flour type/Proofing time (mins.)	Vitamin A level before frying (IU)	Vitamin A level after frying (IU)	% loss
Blank: 45	5000	3000	40
90	3000	2000	33
Treated: 45	24,000	15,000	38
90	27,500	17,000	38
Market: 45	15,000	10,000	33
90	17,000	9000	47

Table 14.
Vitamin A content of fried doughnuts before and after deep frying at 1850C for 5 minutes.

Flour type/Proofing time (mins.)	Iron level before baking (ppm)	Iron level after baking (ppm)	% loss
Blank: 30	32	29	8
60	35	33	6
Treated: 30	54	49	9
60	56	52	7
Market: 30	53	47	11
60	66	61	7

Table 15.
Iron levels of baked pan bread before and after baking at 1750C for 45 minutes.

Flour type/Proofing time (mins.)	Iron level before frying (ppm)	Iron level after frying (ppm)	% loss
Blank: 45	50	47	6
90	41	34	8
Treated: 45	63	57	10
90	78	65	6
Market: 45	60	55	8
90	68	57	15

Table 16.
Iron levels of fried doughnuts before and after deep frying at 185°C for 5 minutes.

Similarly, marked differences are observed in the effects of different ingredients on vitamin A retention in pan bread from fortified flour samples (**Table 17**), as well as under different storage conditions (**Table 18**).

One of the benefits of dough proofing has been reported to be in helping to reduce the level of phytic acid in bread. This is because phytic acid is a known inhibitor of micronutrients present in cereals. Yeast fermentation has also been reported to significantly reduce the phytic acid concentration in bakery products [55], thus increasing the bioavailability of these micronutrients [56, 57]. This is why proofing of dough should be encouraged as a standard practice before baking, deep fat frying, or other

	Flour	Control	High yeast	Low yeast	High salt	Low salt
Treated	26,817	27,066	17,652	21,235	27,282	20,334
Market	10,640	15,207	14,260	15,646	15,950	12,070
Blank	1864	6286	4833	6523	6658	4644

Table 17.
Effect of ingredients on vitamin A retention in pan bread from flour samples (IU).

Day		Sunlight	Shelf	Refrigerator
1	A	1430	1452	1469
	B	27,653	27,830	27,819
	C	11,829	11,935	11,917
2	A	1409	1419	1421
	B	27,624	27,641	27,646
	C	11,812	11,817	11,823
3	A	1392	1408	1414
	B	27,588	27,628	27,637
	C	11,793	11,801	11,814
4	A	1376	1399	1405
	B	27,565	27,613	27,629
	C	11,777	11,794	11,805
5	A	1357	1386	1397
	B	27,526	27,572	27,614
	C	11,754	11,783	11,798

A—Bread from blank flour containing 1000 IU; B—Bread from treated flour containing 30,000 IU of vitamin A/1 kg; C—Bread from standard market sample (fortified).

Table 18.
Effect of storage on vitamin A retention in pan bread from flour sample (IU).

processing methods, as this helps to increase the bioavailability of the iron by reducing the level of phytic acid and its inhibitory effects in the product.

2. Conclusion

Significant advances in the upgrade of traditional processing technologies of roasting, frying, smoking, and fermentation have been recorded in the recent years. The design development and evaluation of a manually operated multi-heat source roaster, which shows significant promise from the evaluation of the proximate, mineral, and vitamin B6 compositions of roasted plantains using gas, electric, or charcoal as the heat sources, has been reported in this review. Consumer perception of the roasted plantain indicated that the product from charcoal roasting was the most acceptable to the panelists. From all indications, the multi-heat source plantain roaster is affordable and could be easily adopted for the purpose of upgrading the rural technology of plantain roasting and eliminating the associated drudgery and likely contamination of such products.

The application of frying technology to traditional deep-fat fried products like *chin-chin* has also been evaluated. Here, it is noted that conductive heat transfer is equal to the combination of convective heat transfer within the hot oil, with conductive heat transfer to the interior of the food and heat losses as a result of radiation. The results show that *chin-chin* fried with corn oil gave the least fat uptake followed by soya oil. This is significant given the knowledge that high fat intake in the human system causes various noncommunicable diseases, especially of a cardiovascular nature [58]. The results also show that vitamin A retention is highest when electric heat is used, compared to other heat sources, with the machine capacity ranging from 5.60 to 8.60 kgh^{-1} depending on the heat source.

From the evaluation of the performance of the smoking kiln for the production of smoked fish and meat products, the efficiency of moisture loss per unit time showed that charcoal is the most efficient as a source of heat by conduction and convection within the smoking chamber as well as by radiation. From the design calculations, it is estimated that the heat capacity of the kiln exceeds the heat requirement for each of the smoked products. The smoking kiln is specifically designed to be relatively cheap, affordable, and easy to maintain for small- and medium-scale processors. The use of fiber glass as an insulator between the double-jacketed walls of the kiln helps to minimize heat losses, thereby increasing efficiency. Consumer perception of various smoked products also showed high acceptability to panelists.

The processing of fermented, dried, and cubed African locust beams has also been shown to be a very significant technology upgrade, with the resulting cubed product being more hygienic and storable, with high nutrient retention compared to the traditional powdered product.

It is concluded therefore that these various advances in the development of affordable, functional, and adaptable technologies offer significant opportunities for small- and medium-scale processors, especially in low- and middle-income countries where adaptability to various energy supply conditions is critical. With the ease of operation, maintenance, and durability, these equipment hold significant promise for use by food processors at small- and medium-scale levels, while with the assurance of high nutrient retention, the benefits of consuming fortified products processed by these traditional technologies will be realizable.

Author details

Olugbenga Ben Ogunmoyela* and Musa Omotayo Jimoh
Bells University of Technology, Ota, Nigeria

*Address all correspondence to: gbenga.moyela2001@gmail.com

References

[1] Sarafa AA, Fausat LK, John KJ, Rowland MO, Olaide AA. Traditional food processing techniques and micronutrients bioavailability of plant and plant-based foods: A review Annals. Food Science and Technology. 2019;**20**(1):30-41

[2] Olayinka K. Upgrading Traditional Food Processing Technologies: The Cassava Example in Nigeria. Food Studies: An Inter-disciplinary Journal. 2014;**3**:1-8

[3] Ogunmoyela OA, Oyewole LO. Improvisation of Equipment for cottage food processing in developing countries. Food Science and Technology Today. Vol 6. Nigeria: Department of Food Science and Technology, University of Agriculture Abeokuta; 1992. pp. 1-6

[4] Oduro-Yeboah C. The role of traditional food processing technologies in preservation of foods: the Ghanaian experience African Journal of Food, Agriculture, Nutrition and Development, 2016;**16**:2. ISSN 1684 5374

[5] Asogwa IS, Okoye JI, Oni K, Promotion of Indigenous Food Preservation and Processing Knowledge and the Challenge of Food Security in Africa. Journal of Food Security. 2017; **5**(3):75-87. DOI: 10.12691/jfs-5-3-3

[6] Ilori TA, Raji AO, Adejumo AO, Kilanko O. Development and Performance Evaluation of a Maize Roaster. International Journal of Science, Technology and Society. 2014;**2**(5): 161-164

[7] Awopetu OO, Aderibigbe AF. Development of a Manually Operated Multi-Purpose Roasting Machine. British Journal of Applied Science & Technology. 2017;**20**(1):1-7

[8] Ogunmoyela OAB, Jimoh MO, Ogabi ON. Development and Evaluation of a Multi-Heat Source Deep Fat Fryer. African Journal of Food Science and Technology. 2016;**7**(3):051-059

[9] Toledo RT. Fundamentals of Food Processing Engineering. 2nd ed. New Delhi, India: CBS Publishers & Distribution; 2000

[10] Sruthi NU, Premjit Y, Pandiselvam R, Kothakota A, Ramesh SV. An overview of conventional and emerging techniques of roasting: Effect on food bioactive signatures, Food Chemistry. Jan 2021;**348**. DOI: 10.1016/j. foodchem.2021.129088. ISSN 0308-8146

[11] Bagheri H, Kashaninejad M, Ziaiifar AM, Aalami M. Novel hybridized infrared-hot air method for roasting of peanut kernels. Innovative Food Science and Emerging Technologies, 2016;**37**: 106-114. DOI: 10.1016/j. ifset.2016.08.014

[12] Eleazu CO, Okafor PN, Ikpeama AI. Total Antioxidant Capacity, Nutritional Composition and Inhibitory Activity of Unripe Plantain (Musa paradisiacae) on Oxidative Stress in Alloxan Induced Diabetic Rabbits. Pakistan Journal of Nutrition. 2010;**9**(11):1052-1057

[13] Idoko JO, Nwajiaku I. Performance of Ripe and Unripe Plantain-Wheat Flour Blend in Biscuit Production. International Journal of Biological Food Veterinary and Agricultural Engineering. 2013;**7**(12):848-851

[14] Oke PK, Ogundare AA. Development of Mechanized Plantain Slicing Machine. Journal of Emerging and Trend in Engineering and Applied Science. 2012;**3**(1):56-60

[15] Jimoh MO, Ogunmoyela OAB, Ogabi ON. Development and

Performance Evaluation of a Multi-Heat Source Plantain Roaster. Nigerian Journal of Technological Development. 2020;**17**(2):120-126. DOI: 10.4314/njtd.v17i2.7

[16] Ezekiel OO, Akinoso R, Olaoye DJ. Design, Fabrication and Evaluation of a Plantain Roaster. Nigerian Food Journal. 2012;**30**(1):131-138

[17] World Health Organization. The World Health Report—Changing History. Geneva: WHO; 2004

[18] Dogan SF, Sahin AC, Sumnu G. Effects of soy and rice flour addition on batter rheology and quality of deep-fat fried chicken nuggets. Journal of Food Engineering. 2005;**71**(1):127-132

[19] Garayo J, Moreira R. Vacuum frying of potato chips. Journal of Food Engineering. 2002;**55**:181-191

[20] Innawong B, Mallikarjunan P, Marcy J, Cundiff J. Pressure conditions and quality of chicken nuggets fried under gaseous nitrogen atmosphere. Journal of Food Processing and Preservation. 2006;**30**:231-245

[21] Ballard T, Mallikarjunan P. The effect of edible coatings and pressure frying using nitrogen gas on the quality of breaded fried chicken nuggets. Journal of Food Science. 2006;**71**(3):259-264

[22] Teye GA, Sheard PR, Whittington FM, Nute SA, Wood JD. Influence of dietary oils and protein level on pork quality, effects on muscle fatty acid composition, carcass, meat and eating quality. Meat Science. 2006;**73**(1):157-165

[23] Mellema M. Mechanism and reduction of fat uptake in deep-fat fried foods. Trends in Food Science & Technology. 2003;**14**:364-373

[24] Shyu S-L, Hau L-B, Hwang LS. Effects of processing conditions on the quality of vacuum-fried carrot chips. Journal of Science, Food and Agric. 2005; **85**:1903-1908

[25] Hussain MA, Rahman MS, Ng CW. Prediction of pore formation (porosity) in foods during drying: generic models by the use of hybrid neural network. Journal of Food Engineering. 2002;**51**:239-248

[26] Devi S, Zhang M, Ju R, Bhandari B. Recent development of innovative methods for efficient frying technology, Critical Reviews in Food Science and Nutrition. 2021;**61**(22):3709-3724, DOI: 10.1080/10408398.2020.1804319

[27] Jimoh MO, Ogunmoyela OAB. Influence of Thermal Properties on Fat Uptake in Deep-Fried Plantain Chips. Journal of Industrial Research and Technology. 2017;**6**(1):63-74

[28] Abiodun MAO, Oluwagbayide SD, Ogunlade CB. Impact of improved smoking kiln design on hygiene and timeliness of drying of smoked fish. Turkish Journal of Agricultural Engineering Research. 2021;**2**(1):133-155

[29] Boye JI, Arcand Y. Current Trends in Green Technologies in Food Production and Processing. Food Engineering Reviews. 2013;**5**:1-17. DOI: 10.1007/s12393-012-9062-z

[30] Dzikunoo J, Letsyo E, Adams Z, Asante-Donyinah D, Dzah CS. Ghana's indigenous food technology: A review of the processing, safety, packaging techniques and advances in food science and technology. Food Control. 2021;**127**:108116. DOI: 10.1016/j.foodcont.2021.108116

[31] Adeyeye SAO. Smoking of fish: A critical review. Journal of Culinary

Science and Technology. 2019;**17**(6):559-575

[32] Macé S, Joffraud JJ, Cardinal M, Malcheva M, Cornet J, Lalanne V, et al. Evaluation of the spoilage potential of bacteria isolated from spoiled raw salmon (Salmo salar) fillets stored under modified atmosphere packaging. International Journal of Food Microbiology. 2013;**160**(3):227-238

[33] Huang C, Shi G. Smoking and microbiome in oral, airway, gut and some systemic diseases. Journal of Translational Medicine. 2019;**17**(1):1-15

[34] Fuentes A, Fernández-Segovia I, Barat JM, Serra JA. Influence of sodium replacement and packaging on quality and shelf life of smoked sea bass (Dicentrarchuslabrax L.). LWT-Food Science and Technology. 2011;**44**(4):917-923

[35] Bashir KMI, Kim JS, An JH, Sohn JH, Choi JS. Natural food additives and preservatives for fish-paste products: A review of the past, present, and future states of research. Journal of Food Quality. 2017;**2017**:1-31

[36] Sutikno LA, Bashir KMI, Kim H, Park Y, Won NE, An JH, et al. Improvement in physicochemical, microbial, and sensory properties of common squid (todarodespacificussteenstrup) by superheated steam roasting in combination with smoking treatment. Journal of Food Quality. 2019;**2019**:1-15

[37] Mohibbullah M, Won NE, Jeon JH, An JH, Park Y, Kim H, et al. Effect of superheated steam roasting with hot smoking treatment on improving physicochemical properties of the adductor muscle of pen shell (Atrina pectinate). Food Science & Nutrition. 2018;**6**(5):1317-1327

[38] Ibarz A, Barbosa-Canovas GV. Unit operation in food engineering. New York, Washington, D.C.: CRC Press; 2003

[39] Tools and basic information for design, engineering and construction of technical application. 2001. Available from: https://www.engineeringtoolbox.com

[40] Rahman MS. Drying of fish and seafood, Handbook of industrial drying. UK: CRC Press; 2006. pp. 547-562

[41] Raji OH, Orelaja OT. Nutritional composition and oil characteristics of golden melon (Cucumismelo) seeds. Food Science and Quality Management. 2014;**27**:2224-6088

[42] Olanbiwoninu AA, Odunfa SA. Microbial interaction in selected fermented vegetable condiments in Nigeria. International Food Research Journal. 2018;**25**(1):439-445

[43] Garcia-Casal MN, Pena-Rosas JPPR, Malavé H. Sauces, spices, and condiments: definitions, potential benefits, consumption patterns, and global markets. Annals of the New York Academy of Sciences. 2016;**1379**:3-16

[44] Asagbra AE, Okafor JWC, Onawola OO, Etoamaihe M, Olatope SOA. Sensory properties of ogiri in Nigerian onugbu soup made from two varieties of melon seeds Cucumis melo and cucumeropsismani. Pakistan Journal of Nutrition. 2012;**11**:596-599

[45] Pathan SB, Senthilkumar R, Meerza AR. Emerging trends in microbial fermentation technologies, Chapter 11. In: Viswanath B, editor. Recent Developments in Applied Microbiology and Biochemistry. India: Academic Press Publishers; 2021.

pp. 113-119. DOI: 10.1016/B978-0-12-821406-0.00011-4

[46] Thompson DE. Design analysis: Mathematical modeling of nonlinear system. UK: Cambridge University Press; 1999

[47] Rahat MA, Raiyan MF, Hossain MD, Ahamed JU, Jony NH. Design and fabrication of a spring constant testing machine and determination of spring constant of a compression spring. International Journal of Engineering Research. 2015;**4**(10):573-577

[48] Shigley JE. Mechanical Engineering Design. NY, USA: McGraw Hill Companies; 1989

[49] Adewumi BA. Modeling the performance of a manually operated legume dehuller using dimensional analysis. Botswana Journal of Technology. 2007;**16**(2):29-36

[50] Oladunmoye MK. Effects of fermentation on nutrient enrichment of locust beans (Parkia biglobosa, Robert bam). Research Journal of Microbiology. 2007;**2**(2):185-189

[51] Zakari UM, Hassan A, Ndife J. Quality Characteristics of African Locust Bean Fruit Pulp Cakes. African Journal of Food Sciences. 2015;**9**(1):17-22

[52] Van Boekel M, Fogliano V, Pellegrini N, Stanton C, Scholz G, Lalljie S, et al. A review on the beneficial aspects of food processing. Molecular Nutrition and Food Research International. 2010;**54**(9): 1215-1247

[53] Ogunmoyela OA, Chigbu LC. Effect of Processing on Vitamin A and Iron Levels in selected fortified foods. Unpublished Results. Ota, Nigeria: Department of Food Science and Technology, Bells University of Technology; 2014

[54] Ogunmoyela OA, Abiodun OA. Evaluation of Vitamin A retention in bread under different processing and storage conditions. Unpublished Results. Ota, Nigeria: Department of Food Science and Technology, Bells University of Technology; 2014

[55] Frontela C, Scarino ML, Ferruzza S, Ros G, Martínez C. Effect of dephytinization on bioavailability of iron, calcium, and zinc from infant cereals assessed in the Caco-2 cell model. World Journal of Gastroenterology. 2009;**15**(6):1977e1984

[56] Almana H. Extent of phytate degradation in breads and various foods consumed in Saudi Arabia. Food Chemistry. 2000;**70**(4):451-456. DOI: 10.1016/s0308-8146(00)00110-2

[57] Leenhardt F, Levrat-Verny MA, Chanliaud E, Remesy C. Moderate decrease of pH by sourdough fermentation is sufficient to reduce phytate content of whole wheat flour through endogenous phytase activity. Journal of Agricultural and Food Chemistry. 2005;**53**:98-102

[58] Hoffman LC, Botha SS, Britz TJ. Sensory properties of hot- deboned ostrich (Struthio camelus var. domesticus) Muscularis gastrocnemius, pars interna. Meat Science and Preservation. 2006;**72**(4):734-740

Chapter 6

Formulation of Edible Bigel with Potential to Trans-Fat Replacement in Food Products

Sunita Mishra and M.A. Firdaus

Abstract

A new issue for the scientific community is to find efficient replacements for unhealthy fat without damaging the organoleptic qualities of the food product in light of growing concerns about the consumption of harmful trans fats in the diet. Bigel is supposedly a novel structured fat system utilised for industrial purposes due to their nutritional advantages, one of numerous solutions intended to replace trans fats in food. These have a lot of potential in the food industry, and are composed of an aqueous phase known as a hydrogel and an organic phase known as an organogel or oleogel. A gel known as an oleogel has oil as its liquid component. Oleogelators, which aid in the development of gels, frequently have low molecular weight, whereas typical hydrogelators have large molecular weight since they are polymeric. A hydrogel is a gel in which water serves as the immobilised phase. Therefore, a bigel is a biphasic system made up of an oleogel and a hydrogel. This chapter will concentrate on the various bigel formulation techniques and chemistry, as well as their latest food uses, and other industries that fit their requirements.

Keywords: bigel, organogel, hydrogel, food applications, trans-fat replacement

1. Introduction

For several food products, fats and oils are important raw materials with some beneficial minerals and vitamins. Saturated fatty acids (SFAs) and trans fatty acids (TFAs), which are present in the form of triacylglycerols, cause the assembly of a colloidal or supracolloidal particle network, which is responsible for structuring the fat into a solid or solid-like substance. This network is largely responsible for the functionality and properties of solid fat [1]. However, solid fats, specifically SFAs and TFAs, raise certain health issues [2]. An increase in low-density lipoprotein cholesterol levels is associated with the consumption of SFAs and TFAs rather than polyunsaturated fatty acids (PUFAs). There is still dispute over the harmful effects of SFAs on health, but if we assume that customers want food free of SFAs, then alternative oil structurants are urgently needed [3]. Since consuming TFAs, a subgroup of unsaturated fatty acids, is associated with increased levels of low-density lipoprotein cholesterol and decreased levels of high-density lipoprotein, they are particularly detrimental [4]. There are few solutions now for removing TFAs while preserving the desired physical characteristics of foods. In this context, the bigel (also known as hybrid gel) is a unique formulation that resembles a solid and is created by combining an oleogel and a hydrogel at a high shear rate [5].

Gels are semisolid mixtures that typically contain two ingredients: liquid and solid, where the solid component is referred to as a gelling agent or gelator and the liquid component is referred to as a solvent. The gelator raises surface tension and is frequently employed at concentrations below 15% w/v to stop solvent flow. Gels are divided into two categories based on the polarity of the solvent: hydrogel and organogel [6, 7]. In contrast to organogels (also known as oleogels), which have apolar liquids as their continuous phase such organic solvents or mineral or vegetable oils, hydrogels are gels whose continuous phase is often a polar solvent, such as water. Cross-linked hydrogels may absorb large amounts of water without dissolving themselves in it. Hydrogels are 3D hydrophilic networks of homopolymeric or heteropolymeric chains. An organic liquid is confined inside a thermoreversible 3D network to create an organogel, a system that resembles a solid. Organogels are extremely simple to make and naturally greasy [8, 9]. Numerous organogelators have been studied, including fatty acids, fatty alcohols, lecithin, physterol and oryzanol mixtures, waxes, steroids, 12-hydroxystearic acid (HSA), L-lysine-based gelators, cyclodextrins, and others. For this type of system, a number of solvents including benzene, hexane, and food oils like sunflower oil, corn oil, sweet almond oil, cod liver oil, and olive oil have also been investigated as liquid phases [10–12].

Bigels' main advantages include the capacity to distribute both hydrophilic and hydrophobic agents, spread easily, be prepared easily, be more stable at room temperature, and be used to control the system's properties by varying the proportion and structural distribution of each phase. Bigels are an appropriate and intriguing formulation for a variety of applications, including medicinal, cosmetic, and food systems [13–15]. These systems can be divided into three categories based on how both phases (organogel and hydrogel) are distributed within bigels: 1. Organogel-in-hydrogel type 2. Hydrogel-in-organogel type 3. Bi-continuous/matrix-in-matrix type. The system that contains organogel as a dispersed phase and hydrogel as a continuous phase is known as an organogel-in-hydrogel system, and it may be the most studied type in the literature. Various researches extensively discussed this kind of bigel. In this bigel type, the oleogel is present as dispersed phase and the hydrogel is present in the continuous phase (**Figure 1**).

This type of bigel has been reported by various researchers by taking into account various hydrogel systems with various hydrogelators/gelling agents, such as gelatin-agar mixture, guar gum, xanthan gum, and acacia gum, gelatin, whey protein, pectin, starch, sodium alginate, sodium carboxy methyl cellulose, hydroxyl propylmethylcellulose, polyvinyl alcohol, and polyvinyl pyrrolidone. Different organogel systems were also taken into consideration to construct bigels in addition to these hydrogels [16–20].

In the last 10 years, a number of researchers have conducted substantial study on bigels, which are developed by combining organogel (oil phase) and hydrogel (aqueous phase) [21, 22]. As a result, this study will focus mostly on this bigel system type. This book chapter's objective is to give readers a solid understanding of various bigel systems. Types and properties of bigels, various bigel synthesis procedures, and various characterisation techniques that can be used to evaluate these systems are the key subjects covered. Additionally, the characterisation techniques of bigel are also discussed using specific instances from the literature as a special system for food applications.

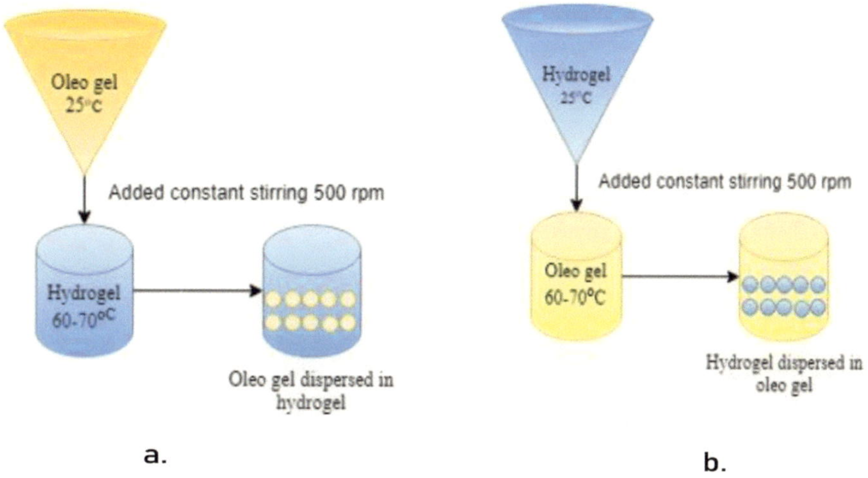

Figure 1.
a. Oleogel dispersed in hydrogel. b. Hydrogel dispersed in oleogel.

2. Preparation of Bigel

Figure 2 illustrates the methods for making bigels by combining hydrogel and organogel. The three most crucial variables for synthesising bigels by combining hydrogel and organogel are the bigels' storage conditions, mixing speed, and temperature. The temperature needs to be held steady during mixing for making bigel. In order to prevent degradation of the hydrogelators during the mixing process, the hydrogelators must be very thermostable. While other study found at mixing the two phases at room temperature while stirring continuously, others reported inclusion of the aqueous phase/hydrogel into the organic phase/organogel at a significantly higher temperature (like 50°C) [23].

The way gels are stored before or after mixing can also have an impact on the final bigels' characteristics. Some bigel systems, as documented in the literature, have been manufactured by storing the final bigel system for a specific amount of time and at a specific temperature after storing the component gels separately for a specific amount of time (e.g., 24 hours). The second strategy, which involves storing the final system after mixing, may result in a more stable system for characterisation. Storing individual gels before combining may assure complete gelling of the individual system [24, 25].

In order to prepare bigels, Satapathy et al. [26] described mixing the individual systems at a somewhat high temperature (50°C), as well as mixing the two phases at room temperature while stirring continuously. Rehman et al. [27] looked at the characteristics of bigels made by combining the two separate systems (hydrogel and organogel) after storing them separately at a specific temperature and time interval. On the other hand, bigels systems can also be made by mixing the various gels before storing the finished product [28]. **Figure 3** depicts the experimental block design for the synthesis of bigels utilising two various approaches. Recent research by Fasolin and Vicente [29] describes how the rheological and microstructural characteristics of bigels are affected by the speed of mixing. To create the bigel system, emulgel/emulsion hydrogel was also combined in various proportions with the organogel phase at room temperature in place of hydrogel. Additionally, a number of models have been

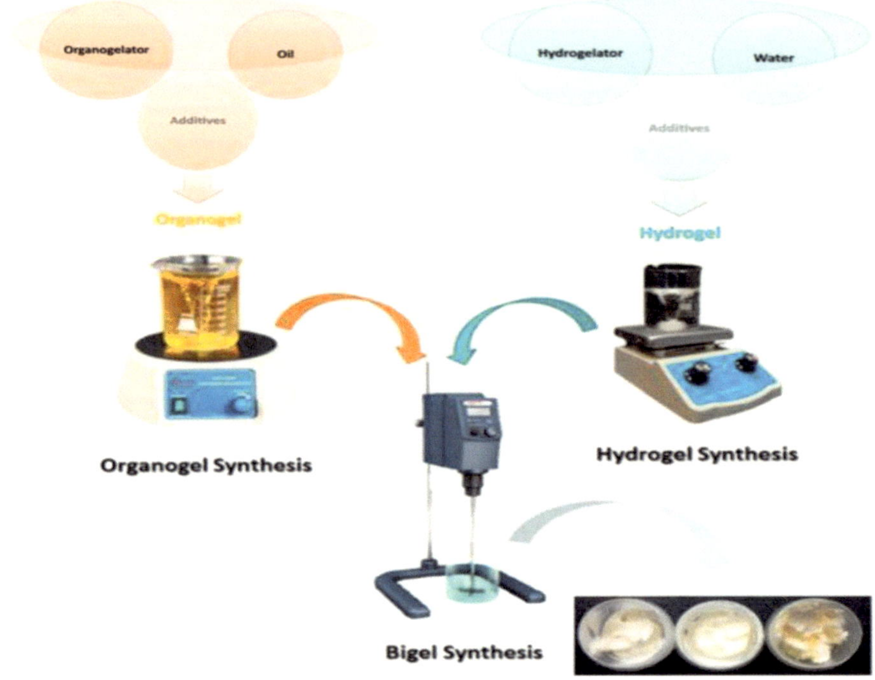

Figure 2.
Schematics of preparation of bigels by mixing hydrogel and organogel.

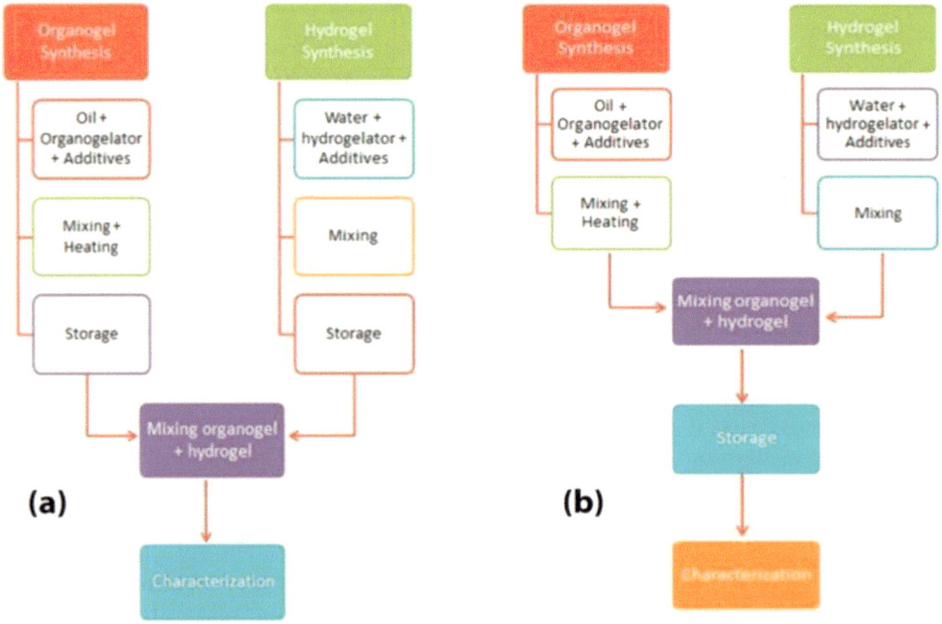

Figure 3.
Flow diagram for synthesis of bigels (a) by storing individual gels before bigel preparation and characterisation (b) by mixing individual phases and then storing bigel formulations before characterisation.

mentioned that can be used to relate the various characteristics of bigels, in particular the rheological models that have recently been proposed in the literature to relate the rheological characteristics of bigels with the dispersed phase fraction as well as with the characteristics of individual phases (organogel and hydrogel) [30, 31].

Lupi et al. [32] discovered olive oil as a solvent and mixture of glyceryl stearate and policosanol as an organogelator, sorbitan monopalmitate-sunflower oil based organogels were investigated by Behera et al. [33], sorbitan monostearate and sesame oil were utilised to prepare organogels as studied by Singh et al. [34], span 60, cetyl alcohol and lecithin-pluronic as organogelators and soya-bean oil as a solvent were found by Ibrahim et al. [7], soya-bean oil-stearic acid based organogels were investigated by Sagiri et al. [35] and Wakhet et al. [36], sorbitan monostearate/medium chain triglyceride based organogels were discovered by Rodrigues et al. [37], sorbitan monostearate as an organogelator and almond oil as a solvent was reported by Andonova et al. [38].

The hydrogel-in-organogel system is a second type of bigel in which the hydrogel phase is dispersed throughout the continuous matrix of the organogel. According to Patel et al. [39], the examination of bigels made by combining organogels made of silica and sunflower oil-fumed organogels at various ratios was done. The results of confocal microscopy supported the morphology of bigels that is of the hydrogel-in-organogel type. The third form of bigel can be thought of as a system with a complex structure where it is challenging to distinguish between the continuous and dispersed phase. A cosmetic system (O/W) was combined with organogels that contained monoglycerides of fatty acids as the organogelator and olive oil as the solvent to create bigels, according to Lupi et al. [30]. The results demonstrated the presence of a complex matrix-in-matrix structure at the highest fraction of organogel.

3. Different formulation methods of edible bigel with potential to replace trans-fat

A novel edible biphasic gel system known as bigel was developed and its mechanical, microstructural, and thermal characteristics was studied by Acevedo & Saffold [40]. The bigel was created using a gelatin hydrogel, a rice bran wax (RBW)-based oleogel, and soybean oil. Bigels were created using three different concentrations of gelatin (5, 7, and 10% (w/w)) and four different oleogel-to-hydrogel (OG:HG) ratios (50,50, 40,60, 30:70, and 20:80). The 10% (w/w) concentration of RBW remained steady. Bigels were examined using confocal laser scanning microscopy (CLSM), small deformation rheology, diferential scanning calorimeter (DSC), and Fourier transform infrared spectroscopy (FTIR). For all bigel formulations, CSLM pictures confirmed an oleogelin-hydrogel system, with an increase in oleogel proportion resulting in larger oleogel droplets and improved stability. All bigel formulations, regardless of gelatin concentration and the ratio of oleogel to hydrogel, exhibited more solid than liquid ($G' > G''$) behaviour and frequency independence at 20°C, according to rheological analysis of the systems. Greater elastic modulus (G') values were consistently seen in bigels with higher OG:HG ratios, such as 50:50 and 40:60, than in those with lower OG:HG ratios, demonstrating that increased oleogel droplet contact results in improved mechanical properties. The Boltzmann Sigmoidal model successfully described the rheological behaviour of all bigels. In all bigel samples, FTIR and DSC analyses revealed discrete peaks for the oleogel and hydrogel phases without any additional thermal events, demonstrating a lack of interactions between the parts of both phases. Overall, the system benefits from having two separate phases and is kinetically stable, making it a "true" bigel.

By combining gelatin hydrogel and stearic acid-based organogel using a heated emulsification technique, Pal et al. [35] created bigel. Sesame oil and soy bean oil based stearate organogels were used to create two distinct types of bigels. Sesame oil and soy bean oil-based gelatin-based emulgels served as the controls. Microscopic examinations showed that the emulgels displayed distributed droplets inside the continuum phase, in contrast to the bigels, which retained droplet clumps. While the interior phase of the bigels barely leached at all, the emulgels revealed a higher level of oil leaching. By using XRD, FTIR, and DSC techniques, the presence of an organogel matrix within the bigels was verified. In comparison to emulgels, bigels demonstrated higher mucoadhesive and mechanical qualities. Studies on cyclic creep-recovery and stress relaxation supported the formulations' viscoelastic properties. The cyclic creep-recovery data were analysed using the four-element Burger's model. Studies on cyclic creep-recovery revealed that the presence of organogels within the structure of bigels may have reduced the amount of deformation. The formulations' nearly complete recovery following the creep stage can be attributed to their increased elastic nature. According to a stress-relaxation study, emulgels had a longer period of relaxation than bigels. Additionally, emulgels had a higher relaxation percentage, indicating a fluid dominance. Human epidermal keratinocyte cell line was used to test the bigels' in vitro biocompatibility (HaCaT). Bigels and emulgels were both biocompatible substances. Based on the findings, it was determined that the created bigels may have a very high potential for usage as emulgel substitutes.

By combining a glycerol monostearate (GMS) oleogel with a gelatin hydrogel with the addition of lecithin and glycerol as surfactant and co-surfactant, respectively, Pinhas et al. [41] developed a unique in-situ bigel system. A bi-continuous system made up of an oleogel crystalline network and a hydrogel polymeric network was created as a result of this combination. On bigel structure, hardness, and stability, the effects of bigel composition, water:oil ratio, and homogenisation time were investigated. Gel hardness and stability were shown to be positively correlated with structuring agent concentration, with the maximum enhancement occurring at 2% weight gelatin and 25% weight GMS. At 5%wt. of the oil phase, lecithin showed maximal strength increase and stabilisation; above that point, a considerable drop was seen. Hardness increased with an increase in oil concentration up to 50%wt, but producing stable bigels was impossible with higher oil contents. When the samples were homogenised for 3 min, better mechanical and stability qualities were seen. With a slight decline in OBC, optimised bigel systems kept their mechanical integrity after 90 days. In addition, compared to the oil sample, low PV, < 10 meqkg-1, was sustained for 30 days at 4°C, 25°C, and 40°C. However, the TBA readings showed a distinct pattern, pointing to a variable transition kinetics between the main and secondary oxidation products. These findings offer a thorough picture of the composition-structure-function relationship of the bigel system and should be taken into account when creating textured food products in the future.

Bigel was created by homogenising at high shear an oleogel emulsion made of soy lecithin, stearic acid, soybean oil, and water, as well as a hydrogel made of whey protein concentrate and water, according to Acevedo et al. [42]. Small angle x-ray scattering, rheology, and fluorescence microscopy were used for characterisation. The oleogel emulsion kept its fundamental structural properties with the addition of the hydrogel component but lost higher order structuring. It was discovered that the bigels' G' values were temperature-dependent. Despite their susceptibility to temperature, the bigels displayed G > G at all ranges from 8 to 98°C. At equal ratios of oleogel emulsion and hydrogel, fluorescence microscopy showed that a bi-continuous bigel was created; however, as either of those phases increased, one of them became the dominant continuous phase. The oleogel emulsion and hydrogel may have interacted to some extent at 10 weight percent water and 15 weight percent protein usage, respectively. This synergy enhanced the mechanical properties of the bigel. On the other hand, the connection between phases changed to be hostile to the mechanical properties of the bigel at protein and water contents outside of those mentioned above.

The objective of the study conducted by Pal et al. [43] was to create and characterise novel bigels for controlled drug delivery applications by combining guar gum hydrogel and sorbitan monostearate-based organogel. The confocal microscopy revealed the presence of a bigel composed of the aqueous and oil phases. Shear-thinning and viscoelastic properties of the bigels were indicated by micro-scale deformation (viscometric) analyses in combination with macro-scale deformation research. Thermal analysis revealed that as the quantity of organogel in the bigels increased, so did their thermal stability.

The bigels that were created had a biocompatible composition. According to an in vitro drug release investigation, the amount of ciprofloxacin (a lipophilic medication) released rose as the amount of organogel content decreased. Further investigation revealed that all of the bigels' drug releases adhered to zero order diffusion kinetics, which is ideal for a controlled release system. The drug-loaded gels effectively combatted Bacillus subtilis' microbes. Finally, the created bigels might be used as matrices for topical medication administration.

In their study, Quilaqueo et al. [44] assessed how the type of hydrogel affected the creation of bigels that would be utilised to substitute fat in cookies. Beeswax/canola oil oleogel, sodium alginate, and carboxymethylcellulose hydrogels were used to make bigels. The outcomes demonstrated that the type of hydrogel utilised had an impact on the peroxide value and binding ability of bigels. They did not change in terms of fatty acid composition, p-anisidine value, oxidative stability, or texture, though. Cookies made with bigels as fat substitutes had a hardness equivalent to that of cookies made with shortening, demonstrating the potential of bigels for usage in food.

4. Characterisation technique

4.1 Organoleptic evaluation

Bigels is left undisturbed once the formulation phase is complete until it reaches room temperature. They are then assessed for a number of factors, including uniformity, colour, pH, viscosity, and phase segregation [30]. Bigel has a high spreading and white colour intensity when the oleogel concentration is high.

4.2 Determination of pH

Digital pH meter are used to determine the pH of the formulation [45].

4.3 Spreadability

The spreadability of the formulation is assessed by layering 0.1 g of gel between two glass slides with the same dimensions (e.g., 25 mm, 1 mm, or 75 mm). Following that, specific weights of 10 g, 20 g, 50 g, or 100 g are loaded for 60 seconds on the upper glass slide. The initial and final spreading diameters are measured before and after each weight is placed [46].

$$\text{Percentage Spreading} = \left[\left(Di - Df\right) / Di\right] \times 100 \tag{1}$$

4.4 Extrudability

A particular quantity of gel is placed into an ointment tube. The length of the gel ribbon that emerged from the ointment tube after uniform pressure was applied is used to calculate the extrudability of the gel [47]. In (cm/s), extrudability is measured.

Extrudability = Distance travelled by gel in cm/10 s

4.5 Thermal analysis

To ascertain the gel-sol transition temperature of organogels, the falling ball method is applied (Tgs). The surface of the organogel is securely supported by a metal ball that weighed around 250 mg. The gel is then heated at a specified rate while a thermometer is inserted, causing the temperature to rise by 1°C per minute-by-minute until it reached 70°C. The temperature of the gel-sol transition is determined as the point at which the ball started to travel through the gel (Tgs). Bigel situations prevent the use of this approach over 50°C because phase separation develops and they become unstable [48].

4.6 Drug content determination

To completely allow for drug leaching, the drug-incorporated bigel is dissolved in phosphate buffer and left undisturbed for at least 48 hours. The drug-containing dispersion is then filtered using Whatmann filter paper. The resulting solution is suitably diluted, and the absorbance is evaluated using a UV spectrophotometer set to the drug's maximum wavelength [49].

4.7 In vitro drug release

A modified Franz diffusion apparatus is used to assess the in vitro release of the medication from the gels using a dialysis membrane (HIMEDIA® LA 330-5MT). A specified quantity of the drug-loaded sample is deposited in the donor compartment, and the phosphate-containing chamber (also known as the receptor chamber) is kept at 32 0.5°C. For 7 hours, a sample of 1 ml is taken every hour, replaced with a new buffer system, and the process repeated. The acquired data is further examined using the Higuchi equation, zero-order, first-order, and Korsmeyer-Peppas models [50].

4.8 Accelerated stability studies

The accelerated stability data are gelled using freeze-thaw (F-T) thermocycles (20 minutes of freezing at 20°C and 20 minutes of thawing at 70°C). The bigels is examined for colour change, viscosity, phase segregation, and homogeneity after each of the five cycles of this procedure. These investigations provide forecasts for long-term stability [51].

4.9 Droplet size distribution

The distribution of droplet size is one of the most important factors in determining how well a medicine is absorbed from gels. Smaller droplet sizes provide us more interfacial surface area, which increases medication absorption and offers great stability. To analyse the droplet size distribution, utilise ImageJ programme [52].

$$\text{Formula used to calculate}: \text{SPAN} = (\text{D90} - \text{D10}) / \text{D50} \tag{2}$$

4.10 Optical microscopy

The mutual arrangement of oleogel and hydrogel within the bigels is investigated using a variety of microscopic techniques, including confocal microscopy, scanning electron microscope (cryo-SEM), and transmission electron microscopy. To distinguish between oleogel and hydrogel, fluorescent dyes are added into both phases [53].

4.11 Fourier transform infrared (FTIR)

The functional groups that are present in the bigel formulation are evaluated using this spectroscopy. Almost majority of the molecules' functional groups absorb infrared light between the wavelengths of 4000 and 1500 cm^{-1}. The spectrum, which was captured in this precise range, is utilised to ascertain if the mixture is lipophilic and hydrophilic. The hydrogel's intra- and intermolecular hydrogen bonding cause a noticeable hump to be present in the range of 3300 to 3200 cm^{-1} [54].

4.12 Mechanical properties

Different approaches are used by viscometers and rheometers to study different mechanical properties of bigels, such as viscosity. The viscosity of the bigels is measured and recorded using a cone and plate viscometer. The measurement is often carried out with a shear rate ranging from 20 to 100 s^{-1} at conventional room temperature (i.e., 25°C). Data are calculated and measured using the Ostwald-de-Waele power law. The viscosity profiles of non-Newtonian fluids are represented by this law.

$\tau = K.\gamma n$ (τ-shear stress, γ-shear rate, K-flow consistency coefficient and n-rate index).

The aqueous and organic phases' interpenetrating network has a synergistic effect on the bigels' rheological properties. Bigel formation will not happen if the hydrogel content in the formulation is greater than 50% of the organogel [55].

4.13 Electric conductivity

The electrical characteristics of the bigels are examined in order to appreciate the conductivity profiles. The electrical profiles of the bigel can be calculated using a computer-controlled impedance analyser. A specified frequency range is used to measure the data while it is at room temperature. We are better able to comprehend their transport behaviour when we consider how the current affects the formulation's conductivity. The biggerels with a higher fraction of hydrogel display stronger conductivity because of the ions that are present in the water phase [56].

4.14 Photostability study

To evaluate the stability profile of the bigels in the presence of light, photostability investigations are carried out. This further aids us in choosing the kind of packaging that will be applied to the finished commodity (i.e Primary or secondary packaging). The ICH Q1B guideline has been followed in the execution of these research. The product is tested under four distinct conditions: first, it is wrapped in aluminium foil; second, it is packed in cardboard; third, it is kept in a container with a label; and fourth, it is kept in a container without a label. The formulation is assessed for different characteristics, including pH, contaminants, assay, and appearance after exposure to a certain amount of light [57].

5. Conclusion and future perspective

Different bigel systems have recently been created and modified to suit the requirements of various applications. This book chapter discussed the modelling of these systems in detail and presented the literature on the significant bigel properties. Additionally, it has been considered to use these systems for food applications. Bigels are a new class of materials, hence thorough analysis of these systems is necessary for applications in industry. Storage of bigels, mixing speed, mixing temperature, the amount of hydrogelator and organogelator used, the ratio of hydrogel to organogel, the addition of emulsifiers, and the use of emulgels in place of either organogel or hydrogel are some of the variables that are crucial in the synthesis of bigels. Researchers need to focus further on how the aforementioned criteria affect the final attributes of the created formulation. While the hydrogel-in-organogel and bi-continuous forms of bigels have received less attention from researchers, the organogel-in-hydrogel type has been extensively examined in the literature. In order to better propose a model to precisely anticipate the features of such a complicated system, more research is required to comprehend the reliance of the rheological properties of bigels on the moduli of dispersed and continuous phase as well as on the particle size distribution. Additionally, such systems can be simulated to compare the outcomes with real-time responses. Furthermore, bigel formulations can be created and examined for a wide range of culinary applications in the future, in addition to medicinal and cosmetic applications.

Funding

This research received no external funding.

Conflict of interest

The authors declare that there is no conflict of interests among them regarding the publication of this paper.

DOI: http://dx.doi.org/10.5772/intechopen.110517

Author details

Sunita Mishra* and M.A. Firdaus
Department of Food and Nutrition, School of Home Science, Babasaheb Bhimrao Ambedkar, A Central University, Lucknow, India

*Address all correspondence to: sunitabbau@gmail.com

References

[1] Co E, Marangoni AGO. An introduction. In: Marangoni A, Garti N, editors. Edible Oleogels Structure and Health Implications. 2nd ed. Cambridge, MA, USA, Urbana, IL, USA: Academic Press; AOCS Press; 2018. pp. 1-30

[2] Firdaus MA, Sunita M. Nutritional recommendations for management of coronary heart disease during viral infections with an emphasis on COVID-19. Journal of Cardiovascular Disease Research. 2021;**12**:982-991

[3] Ahmad S, Ujala F, Tanveer I. Key characteristics and modelling of bigel systems: A review. Materials Science & Engineering C. 2019;**97**:932-953

[4] Firdaus MA, Mishra S. Industrial trans-fatty acid intake associated with coronary heart disease risk: A review. Journal of Nutrition & Food Sciences. 2021;**11**:806

[5] Ahmad S, Francesca RL, Domenico G. Bigels: A unique class of materials for drug delivery applications. Soft materials. 2018;**16**:1-30

[6] Kamaly N, Yameen B. Degradable controlled release polymers and polymeric nanoparticles: Mechanisms of controlling drug release. Chemical Reviews. 2016;**116**(4):2602-2663

[7] Mohmoud MI, Salma AH, Mahmoud MM. Organogels, hydrogels and bigels as transdermal delivery systems for diltiazem hydrochloride. Asian Journal of Pharmaceutical Sciences. 2013;**8**:48-57

[8] Shakeel A, Farooq U, Iqbal T, Yasin S, Lupi FR, Gabriele D. Key characteristics and modelling of bigels systems: A review. Materials Science and Engineering: C. 2019;**97**:932-953

[9] Velichka A, Petya P, George SG, Vencislava TT. Ketoprofenloaded polymer carriers in bigel formulation: An approach to enhancing drug photostability in topical application forms. International Journal of Nanomedicine. 2017;**12**:6221-6238

[10] Sawalha H, Venema P, Bot A, Flöter E, van der Linden E. Food Biophysics. 2011;**6**:20-25

[11] Mallia VA, George M, Blair DL, Weiss RG. Langmuir. 2009;**25**:8615-8625

[12] Suzuki M, Nanbu M, Yumoto M, Shirai H, Hanabusa K. New Journal of Chemistry. 2005;**29**:1439-1444

[13] Behera B, Dey S, Sharma V, Pal K. Rheological and viscoelastic properties of novel sunflower oil-span 40-biopolymer–based Bigels and their role as a functional material in the delivery of antimicrobial agents. Advances in Polymer Technology. 2015;**34**(2):21488

[14] Almeida I, Fernandes A, Fernandes L, Pena Ferreira M, Costa P, Bahia M. Pharmaceutical Development and Technology. 2008;**13**:487-494

[15] Lupi F, Gentile L, Gabriele D, Mazzulla S, Baldino N, De Cindio B. Journal of Colloid and Interface Science. 2015;**459**:70-78

[16] Velichka YA, Petya TP, Elisaveta GA, Teodora DD. Carbopol hydrogel/sorbitan monosterate- almond oil based organogel biphasic formulations: Preparation and characterization of the bigels. Tropical Journal of Research. 2017;**16**(7):1455-1463

[17] Narayana CR, Arjun M, Sandeep DS. Design and evaluation of bigels containing flurbiprofen. Research Journal of Pharmacy and Technology. 2018;**11**(1):143-152

[18] Martins AJ, Silva PMPD, Maciel JFG, Pastrana L, Cunha RL, Cerqueira MA, et al. The 19th Gums & Stabilisers for the Food Industry Conference: Hydrocolloid Multifunctionality. Berlin, Germany: RSC Publishing; 2017

[19] Paul SR, Qureshi D, Yogalakshmi Y, Nayak SK, Singh VK, Syed I, et al. Journal of Surfactants and Detergents. 2018;**21**:17-29

[20] Hamed R, Abu Rezeq AA, Tarawneh O. Development of hydrogels, oleogels, and bigels as local drug delivery systems for periodontitis. Drug Development and Industrial Pharmacy. 2018;**44**:1488-1497

[21] Deng G, Ma Q, Yu H, Zhang Y, Yan Z, Liu F, et al. Macroscopic Organohydrogel hybrid from rapid adhesion between dynamic covalent hydrogel and Organogel. ACS Macro Letters. 2015;**4**(4):467-471

[22] Blumlein A, McManus JJJJOMCB. Bigels formed via spinodal decomposition of unfolded protein. Journal of Materials Chemistry. 2015;**3**(17):3429-3435

[23] Di Michele L, Varrato F, Kotar J, Nathan SH, Foffi G, Eiser E. Multistep kinetic selfassembly of DNA-coated colloids. Nature Communications. 2013;**1302**:5559

[24] Gabriele D, de Cindio B, D'Antona P. Rheologica Acta. 2001;**40**:120-127

[25] Charyulu NR, Muaralidharan A, Sandeep D. Design and evaluation of bigels containing flurbiprofen. Research Journal of Pharmacy and Technology. 2018;**11**:143-152

[26] Satapathy S, Singh VK, Sagiri SS, Agarwal T, Banerjee I, Bhattacharya MK. Development and characterization of biphasic gels for potential application as semi-solid fat replacers in animal- and plant-based food products. Journal of Applied Polymer Science. 2015;**132**:476-481

[27] Rehman K, Amin MCIM, Zulfakar MH. Development and physical characterization of polymer-fish oil bigel (hydrogel/oleogel) system as a transdermal drug delivery vehicle. Journal of Oleo Science. 2014;**63**:961-970

[28] Yu G, Yan X, Han C, Huang F. Characterization of supramolecular gels. Chemical Society Reviews. 2013;**37**(16):6697-6722

[29] Fasolin LH, Vicente A. The 19th Gums & Stabilisers for the Food Industry Conference: Hydrocolloid Multifunctionality. Berlin, Germany: RSC Publishing; 2017

[30] Kodela SP, Pandey PM, Nayak SK, Uvanesh K, Anis A, Pal K. Novel agar–stearyl alcohol oleogel-based bigels as structured delivery vehicles. International Journal of Polymeric Materials and Polymeric Biomaterials. 2017;**66**:669-678

[31] Craig DQ, Reading M. Thermal Analysis of Pharmaceuticals. Boca Raton: CRC Press; 2006

[32] Lupi FR, Shakeel A, Greco V, Rossi CO, Baldino N, Gabriele D. A rheological and microstructural characterisation of bigels for cosmetic and pharmaceutical uses. Materials Science and Engineering C. 2016;**69**:358-365

[33] Behera B, Sagiri SS, Singh VK, Pal K, Anis A. Mechanical properties and delivery of drug/probiotics from starch and non-starch based novel bigels: A comparative study. Starch-Starke. 2014;**66**(9-10):865-879

[34] Singh VK, Anis A, Banerjee I, Pramanik K, Bhattacharya MK, Pal K.

Preparation and characterization of novel carbopol based bigels for topical delivery of metronidazole for the treatment of bacterial vaginosis. Materials Science and Engineering: C. 2014;**44**:151-158

[35] Sagiri SS, Singh VK, Kulanthaivel S, Banerjee I, Basak P, Battachrya MK, et al. Stearate organogel-gelatin hydrogel based bigels: Physicochemical, thermal, mechanical characterizations and in vitro drug delivery applications. Journal of the Mechanical Behavior of Biomedical Materials. 2015;**43**:1-17

[36] Wakhet S, Singh VK, Sahoo S, Sagiri SS, Kulanthaivel S, Bhattacharya MK, et al. Characterization of gelatin-agar based phase separated hydrogel, emulgel and bigel: A comparative study. Journal of Materials Science: Materials in Medicine. 2015;**26**(2):118

[37] Rodrigues SF, Fiel LA, Shimada AL, Pereira NR, Guterres SS, Pohlmann AR, et al. Lipid-Core Nanocapsules act as a drug shuttle through the blood brain barrier and reduce glioblastoma after intravenous or Oral administration. Journal of Biomedical Nanotechnology. 2016;**12**(5):986-1000

[38] Andonova VY, Peneva PT, Apostolova EG, Dimcheva TD, Peychev ZL, Kassarova MI. Carbopol hydrogel/sorbitan monostearate-almond oil based organogel biphasic formulations: Preparation and characterization of the bigels. Tropical Journal of Pharmaceutical Research. 2017;**16**(7):1455-1463

[39] Patel AR, Mankoc B, Sintang MDB, Lesaffer A, Dewettinck K. Fumed silica-based organogels and 'aqueous-organic' bigels. RSC Advances. 2015;**5**(13):9703-9708

[40] Saffold AC, Acevedo NC. Development of novel Rice bran wax/gelatin-based biphasic edible gels and characterization of their microstructural, thermal, and mechanical properties. Food and Bioprocess Technology. 2021;**14**(12):2219-2230

[41] Samui T, Goldenisky D, Rosen-Kligvasser J, Davidovich-Pinhas M. The development and characterization of novel in-situ bigel formulation. Food Hydrocolloids. 2021;**113**:106416

[42] Bollom MA, Clark S, Acevedo NC. Development and characterization of a novel soy lecithin-stearic acid and whey protein concentrate bigel system for potential edible applications. Food Hydrocolloids. 2020;**101**:105570

[43] Singh VK, Banerjee I, Agarwal T, Pramanik K, Bhattacharya MK, Pal K. Guar gum and sesame oil based novel bigels for controlled drug delivery. Colloids and Surfaces B: Biointerfaces. 2014;*123*:582-592

[44] Patel AR, Nicholson RA, Marangoni AG. Applications of fat mimetics for the replacement of saturated and hydrogenated fat in food products. Current Opinion in Food Science. 2020;*33*:61-68

[45] Banks SR et al. The influence of substituted phenols on the sol: Gel transition of hydroxypropyl methylcellulose (HPMC). Aqueous Solutions. 2014;**101**:1198-1204

[46] Aejaz A et al. Formulation and in vitro evaluation of aceclofenac solid dispersion incorporated gels. 2010;**2**:1

[47] Mohanan S, Rasheed N, Bimal Raj KS. Formulation and evaluation of antimicrobial gels for the treatment of paronychia. International Journal of Applied Pharmaceutics. 2018;**10**(6):161-167

[48] Mazurkeviciute A, Ramanauskiene K, Ivaskiene M, Grigonis A, Briedis V. Acta Pharmaceutica. 2018;**68**:223-233

[49] Balata G, Shamardl HJJOIIP, Sciences B. Formulation of chlorpheniramine maleate in span 60/tween 20 based organogels for transdermal delivery. Journal of Sci & Engineering. 2017;**4**(1):49-57

[50] Venkateswara Rao S, Vijaya Sri P, Padmalatha K. Formulation and evaluation of oxiconazole emulgel for topical drug delivery. 2017;**9**(10):58876-58878

[51] Singh VK et al. Bigels. In: Polymeric Gels. India: Woodhead Publishing; 2018. pp. 265-282

[52] Yan Y-D et al. Enhanced oral bioavailability of curcumin via a solid lipid-based self- emulsifying drug delivery system using a spray-drying technique. Biological and Pharmaceutical Bulletin. 2011;**34**(8):1179-1186

[53] Pal R. Electromagnetic, Mechanical, and Transport Properties of Composite Materials. Boca Raton: CRC Press; 2014

[54] Rehman K, Zulfakar MH. Novel fish oil-based Bigel system for controlled drug delivery and its influence on immunomodulatory activity of Imiquimod against skin cancer. Pharmaceutical Research. 2017;**34**(1):36-48

[55] Neves JD et al. Rheological properties of vaginal hydrophilic polymer gels. Current Drug Delivery. 2009;**6**(1):83-92

[56] Witt JA, Mumm DR, Mohraz A. Bijel reinforcement by droplet bridging: A route to bicontinuous materials with large domains. Soft Matter. 2013;**9**(29):6773-6780

[57] Guideline, I.H.T.J.Q.B. Current step, stability testing: Photostability testing of new drug substances and products. In: International Conference on Harmonisation of Technical Requirements for Registration of Pharmaceuticals for Human Use. Vol. 41996